SHIA SU

ZERO WASTE

SIMPLE LIFE HACKS TO DRASTICALLY REDUCE YOUR TRASH

Skyhorse Publishing

Printed on sustainably sourced paper using vegetable inks.

Skyhorse Publishing books may be purchased in bulk at special discounts for sales promotion, corporate gifts, fund-raising, or educational purposes. Special editions can also be created to specifications. For details, contact the Special Sales Department, Skyhorse Publishing, 307 West 36th Street, 11th Floor, New York, NY 10018 or info@skyhorsepublishing.com.

Visit our website at www.skyhorsepublishing.com.

10 9 8 7 6 5 4 3 2

Library of Congress Cataloging-in-Publication Data is available on file.

Cover design and interior layout by Christina Diwold
Cover photo credit by Shia Su
All photos by Shia Su, unless mentioned otherwise.
Photos of nuts on pages 111, 112, and 113 from Freedom_Studio, Binh Thanh Bui, Tukaram.Karve, and windu/Shutterstock.com.

Print ISBN: 978-1-5107-3081-6
Ebook ISBN: 978-1-5107-3082-3

Printed in China

For my mom.

TABLE OF CONTENTS

OLIVA
MAYATI

INTRODUCTION

ZERO WASTE LIVING–
A MODERN REEMERGENCE
OF AN OLD LIFESTYLE

Many people find it hard to wrap their heads around how little trash my partner and I produce at home. "No way, that's unreal!" is the most common reaction we get when people find out that our annual collection of nonrecyclable and plastic waste amounts to a quart jar.

Well, we do not live completely *zero* waste, as our waste-stuffed jar proves. And yes, there is more: 6.5 pounds of paper waste, 0.2 pounds of metal waste such as bottle caps and staples, a dozen bottles and jars, and kitchen scraps (which we now compost in our kitchen after we finally gathered enough courage to give the rather unconventional idea of having worms as pets a try [see p. 155]). All this waste is recycled; our jar, on the other hand, contains nonrecyclable materials (plastic is recyclable in theory, but it mostly fails to be recycled for a wide range of reasons [see p. 150]).

Whenever we talk to older folks, we get a very different reaction. They usually laugh out loud, saying to us "youngsters": *Oh, come on! Living waste-free is old news!* At this point, they often start to give us practical advice: how to unclog the sink without all the chemical gimmicks, for example; and have we ever tried to just wrap our sandwich in a dish towel?

As a matter of fact, only a few decades ago, everybody lived a "zero waste lifestyle." Of course, nobody called it "zero waste" back then—it was just the *normal* way of life. We have only adapted a wasteful lifestyle very recently, and one can argue that this is not progress but rather a short-sighted high with a sudden crash that will be served with a side of regret.

WE ARE COLLECTIVELY BLIND TO TRASH

Trash has become such an integral and seemingly natural part of our daily lives that we never stop to think about it. We toss the empty shampoo bottle into the trash can. We bring out the trash. The moment the smelly bag leaves our home, it is out of sight, out of mind.

Of course, we know that this trash does not simply vanish into thin air. We know about toxic landfills. Some of us already know that recycling isn't as green as we would like to think—shipping recyclables across the globe is a common practice. We have heard about our oceans drowning in plastic. We are concerned that trash is disrupting the entire food chain and already found its way onto our plates.[1] But for some reason, none of that knowledge is at the forefront of our minds when we shop, grab a coffee, or are freeing an organic cucumber from its plastic shrink-wrap.

According to the United States Environmental Protection Agency, the average American generated 1,620 pounds of garbage in 2014. This is a whopping 4.44 pounds per day! But what happens to all of it after it has been collected? In the US, only 34.6 percent is recycled, 12.8 percent is combusted for energy recovery, and the rest is landfilled.[2] As a comparison, several European countries like the Netherlands, Germany, or Sweden have already banned the practice of landfilling garbage.[3]

Waste management facility.

Around the globe, countless campaigns have been run to promote recycling. However, in the end, recycling only fights the symptoms, but not the cause, of our waste problem, especially since we often take "recyclable" labels as a permission to produce waste—"It's OK, it's recyclable!" Wouldn't it be better not to trash the environment in the first place and only attempt to fix the damage *afterward*? Not to mention that what we commonly refer to as "recycling" is in fact often only "down-cycling"—the transformation of the original material into a lower quality product that is oftentimes no longer recyclable.

If we are serious about counteracting climate change, the only way to make a difference is to reduce our impact on the environment and reframe our mind-set.

It is not about recycling more. It is about wasting less.

---

1 Weikle, "Microplastics found in supermarket fish, shellfish," or Smillie, "From sea to plate."

2 United States Environmental Protection Agency, "Advancing Sustainable Materials Management: 2014 Fact Sheet"

3 European Environment Agency, "Municipal waste management across European countries."

A "ZOMBIE APOCALYPSE-KIND OF MISTAKE"?

As we all know too well, our economy is based on growth. This means we must consume more and more in order to keep hitting those growth expectations that signal a "healthy" economy. And we do a pretty good job at keeping up. We continue buying things, even though we already have an abundance of stuff at home, stuff that is much more than we need or could ever use. **If everyone on this planet consumed as much as the average US citizen, we would need four earths to sustain us!**

Too bad we only have this one planet with only a limited amount of resources. Unlimited growth is simply impossible to sustain in the long term, yet it is a system by which we still operate. I argue that a system based on unlimited growth was set up to fail from the very beginning. I like to call it a "zombie apocalypse–kind of mistake"—a thoughtless error with a fatal outcome.

Things need to change, whether we accept the facts or not. Some economists call for a focus on qualitative instead of quantitative growth, while degrowth theorists envision alternative social structures to achieve economic equality with the limited resources available to us. They plead for "better" instead of "more."

> "We do not like to hear that the party is over–especially when we are privileged enough to live in a rich society."
>
> —Harald Welzer, social psychologist, Europa-Universität Flensburg[4]

4 Norddeutscher Rundfunk, "Neuland."

Moreover, satisfying our greed for faster, cheaper, and newer does not even seem to make us happy! In fact, the national happiness in the US peaked sometime in the 1950s,[5] i.e., it only went downhill from then on. What is for sure is the fact that our hunger for *more* results in misery on the other side of the globe, where laborers (some of them children) are exploited to satisfy our cravings for fast fashion and cheap commodities.

Zero waste is about reducing waste first and foremost. However, when you start to minimize the trash you create, one side effect is you will automatically begin to buy less; you will also buy more consciously, focusing on what truly makes you *happy*, rather than a quick high. You'll start to focus on *better* instead of more.

Designed to be trashed.

Remember: the natural resources available to us are limited and running low. Yet we are treating them like income instead of irreplaceable capital, as economist E. F. Schumacher pointed out in 1975.[6] We are wasting an enormous amount of scarce and limited resources like fossil fuel on producing disposable plastic items. **Isn't it crazy how we are wasting our dwindling resources on things designed to go directly into the trash after one use?**

Furthermore, as packaging and single-use items are designed to only be utilized once, they need to be cheap. To cut costs, they must be produced inexpensively. This is only achievable by externalizing the cost, which in turn means that the workers and the environment pay for the low prices. Once-valuable resources are often converted to toxic waste, which finds its way back to us by entering the food chain (e.g., microplastics) or by polluting our groundwater. Ultimately, we are also paying for the cheap price tags with our health.

5 McKibben, *Deep Economy*, 35–36.
6 Schumacher, *Small is Beautiful*, 14.

BE PART OF THE SOLUTION, NOT THE PROBLEM

Personally, I have grown so tired of the blame game. The industry blames the consumer for only accepting perfect goods at ridiculously low prices, preventing them from offering more sustainable options. At the same time, consumers say it is the industry that makes large-scale decisions, so, duh, it is the industry that needs to change first.

It is a ping-pong game without a winner, and no one takes responsibility. It is easy to feel overwhelmed, to feel that you are only one in a million who cannot make large-scale decisions. I cannot pass any laws, nor am I the CEO of a big corporation. But does this make me powerless? It does not. I for one might not be able to have a direct influence on the upstream waste created in the production and transportation processes. However, I *can* set an example by refusing waste at the level of the consumer.

To me, being zero waste means starting with myself and what I have direct control over.

I realized I have full control over my own consumption! It was about time I turned my back on this consumption craze and put my money where it would do good instead.

People like to tell me that what I do is nothing more than a drop in the ocean. But try to think about it another way. Climate change has not been, and could have never been, caused by a single person who called all the shots; it is a result of millions of people, contributing over time, one by one. One of my coworkers used to say: "You need to climb down the ladder the way you climbed it up: step by step." It goes without saying that I would rather make a positive impact than a negative one.

There are many little things we can do to contribute. Every time we spend money on something, we create a demand and vote for more of it to be produced, whether it is our intention or not. If I buy fast fashion, more fast fashion will be produced. Our purchases tell companies that it is lucrative to continue to produce the way they do.

The good news is this also means we have the power to support businesses that do work hard to make a difference. However, note that if we choose not to be part of the system—for example, by going off the grid or becoming freegan (living off only what would otherwise be wasted) —we lose our vote. While being freegan has a true minimal impact, I personally choose to generate demand for things that I believe should one day become the norm instead of the exception: organic produce, fair trade goods, and toxin-free products that are produced with minimal waste and packaging.

Generating demand for more sustainable options can be quite the game changer. Even though there is always room for improvement, these businesses put pressure on the big players to catch up—and this is how the rules of the economic system can change little by little toward a better environment for all of us.

The Soap Dispensary, Vancouver, BC, Canada.
Photo credit: Fahim Kassam

The Soap Dispensary, Vancouver, BC, Canada.
Photo credit: Hanno Su

Zero waste bulk stores strive to enable their customers to live zero waste lifestyles.

Maybe it is time to stop defining ourselves as "consumers" but as members of a community. A member of a community strives for "better," not "more." When we stop buying more, we will suddenly find ourselves with time on our hands for the important things in life. By buying only what we truly need, we save a lot of money—which in turn allows us to afford safer, healthier, and fairer options, like organic produce!

ZERO WASTE LIFESTYLE

BENEFITS OF A **ZERO WASTE LIFESTYLE**

LET'S BE HONEST

Most of us would not trade personal comfort for the "greater good." I am afraid my husband, Hanno, and I are no exceptions. Each and every year, we always come up with the most ambitious resolutions, but we rarely stick to any of them.

Sometimes, I feel a bit like a fraud when people praise me for my efforts. To be perfectly honest, I have a hard time not hitting the snooze button in the morning (I know, I know). Yet we stuck to a zero waste lifestyle despite being notoriously easily distracted. **It is perhaps more accurate to say that we slipped down the zero waste rabbit hole because of all the benefits that come with such a lifestyle that we wouldn't want to miss for the world!**

BE HEALTHIER

Despite the fact that plastics leach harmful substances, like bisphenol A (BPA) and phthalates, almost everything comes wrapped in a layer of plastic nowadays—be it a cucumber, shampoo bottle, or even a carton of water. These substances are suspected causes of cancer, early puberty in girls, infertility in males, hyperactivity, and neurological conditions. They are also linked to obesity and type 2 diabetes. I found it shocking to learn that BPA is commonly traceable in urine, according to an article in the Canadian Press.[7]

Single-use plastics, brand-new plastic items, and new clothes are especially prone to leaching an alarming amount of harmful substances and

7 The Canadian Press, "Most Canadians have BPA in urine, lead traces in blood."

toxins. Becoming zero waste reduces your exposure to these toxins tremendously.

Aggressive chemicals in cleaning products and the cocktail of chemical compounds in cosmetics will cease to be an issue as well when you live zero waste—you will most definitely start using natural remedies instead (if you suffer from atopic eczema, like I do, this will be a great relief). You will also gradually replace processed and junk food with nutritious, real, and even organic food.

SAVE MONEY

At first glance, a zero waste lifestyle seems elitist. Going to the farmers market or buying organic is indeed more costly. However, in total, our expenditure is a lot less compared to before we went zero waste, even though we only buy organic produce today.

We save more money in other areas than the amount we spend on organic food. It makes sense, too. Entire categories—for example, pretty much all drugstore items—simply disappeared from our budget list. Here are a few things we learned.

› **We consume less.** We only buy what we need, and we usually only buy to *replace* and not *add* to our inventory. Whatever impulse buys we allow ourselves to indulge in also become a lot more wallet-friendly—instead of buying a piece of clothing on impulse, we buy broccoli that was not originally on our shopping list.

› **Many conventional items are a lot more expensive than they appear.** Cleaning products, cosmetics, and makeup are actually quite expensive! We have simply gotten used to the price tag. Processed food is more expensive than preparing the very same thing from scratch. Junk food will cost you in the long run in the form of health issues and doctor's bills.

- **Quality over quantity.** There is a German proverb that says, "Whoever buys cheap pays twice." Buying high-quality items that are made to last a lifetime might be more expensive at first, but it will pay off in the long run.

- **Reusables instead of single-use items.** Single-use items like wipes, tissues, paper towels, tinfoil, or wax paper are designed to be tossed. This means we have to keep buying them over and over again our entire lives. Swapping these for reusable alternatives will save money in the long run.

- **Go for tap water.** Did you know that bottled water can cost up to five hundred times more than tap water (based on your utility bills), even though it is less regulated? Not to mention that the plastic bottle has leached BPA or phthalates into the water by the time you are drinking it.[8] Hard water is, in fact, water with a high mineral content. You can easily find out if you live in a state or area that has hard water by checking your local water supplier's website. Your appliances might not like hard water, but you should. In fact, hard water is, by definition, water rich in minerals. Drinking hard water contributes to our calcium and magnesium needs, as the United States Geological Survey points out.[9]

- **Less (stuff) is more (money).** Having a lot of belongings can be costly. We need to store, maintain, and repair it all. If you are renting a storage unit, moving houses, or buying more items to organize and accommodate your stuff, you are probably very aware of the fact that ownership costs hard-earned money.

- **Pay less for trash collection.** Many US municipalities charge by the amount of trash they have to pick up. Get a smaller trash can or use fewer bags and save money.

---

8 Carwile et al., "Polycarbonate Bottle Use and Urinary Bisphenol A Concentrations."
9 United States Geological Survey, "Water Hardness."

TRACES OF LEAD IN HOUSEHOLD PLUMBING

According to the American Water Works Association, 6.5 million lead pipes are still in use—and not only in old buildings. Many cities and areas across the United States still use old lead service lines. Some cities, like New York City, treat their water to reduce the amount of lead that dissolves into the water by adding phosphoric acid and monitoring and adjusting the pH level. If you live in a building built before 1985, you can easily check if your pipes contain lead. If the pipes are dark grey, can be scratched easily, and the scratch looks silvery—it is a lead pipe.

Lead in drinking water should *not* be taken lightly. Boiling the water does *not* remove lead but can concentrate lead levels instead. Ask your local water department if they still use lead service lines. Check the water pipes in your home if you live in an older building. You are most likely fine, but if your tap water does contain lead, consider using a water filter that is certified to be able to remove lead in accordance to the standards developed by the National Sanitation Foundation (NSF International).

SIMPLIFY YOUR LIFE!

Everywhere we go, we are confronted with a sheer overwhelming variety of *everything*, which encourages us to consume more, and very successfully at that, I may add. We all believe we need highly specialized one-trick ponies on steroids for the most elementary household tasks.

In 2008, the average supermarket carried almost 47,000 products. That is more than five times that of what a supermarket in 1975 used to offer.[10]

How does this selection of products affect us on a psychological level? Because we are overwhelmed, the decision-making process is stressful. This is called "decision fatigue." Why would you want to waste your mental capacities on making mundane decisions?

In contrast, zero waste helps to narrow down your choices and allows you to enjoy a clutter-free mind in these busy times.

MORE TIME FOR WHAT REALLY MATTERS

Whenever I tell people that I buy everything in bulk, they assume that we must go to at least twenty-seven different stores to get all the groceries we need for the week. Luckily, our reality could not be more different. In fact, buying groceries is something we do not worry about anymore (see p. 55)!

10 Consumer Reports, "What to Do when There Are Too Many Product Choices on the Store Shelves?"

West Coast Refill, Victoria, BC, Canada.

The same goes for trash. Thankfully, trash has ceased to consume a large part of our daily lives—no more figuring out how to recycle a certain type of plastic, no more junk mail to deal with, no more climbing stairs to bring out the trash from the fifth floor.

Decluttering has left our entire home looking tidy and neat. Household chores have become a breeze.

Instead of spending our time in stores, fulfilling our roles as consumers (earning money to spend money), sorting recyclable items, organizing our belongings, or cleaning our home, we now enjoy spending more time with loved ones, tackling those if-we-just-had-the-time-to-do-it projects, or just lazing around and being silly.

CONSUME CONSCIOUSLY, FEEL EMPOWERED

In 2013, total spending on television advertisement reached $78 billion. TV commercials have become shorter, cramming more into each advertisement block.[11] This is one and a half times what the Congressional Budget Office estimated the US government would spend on health insurance subsidies in 2017![12] We, as a society, seem to place more importance on getting people to consume more than on saving lives.

11 Luckerson, "Here's Exactly Why Watching TV Has Gotten So Annoying."
12 Congressional Budget Office, "Health Care."

We, as "consumers," are ultimately the ones paying the costs of production and the airing of commercials. After all, advertisements are designed to rake in more business and bind us to a certain brand. I don't know about you, but that is not where I want my hard-earned money to go; neither do I want to pay for the packaging products come in.

Once we started living zero waste, I learned to turn a blind eye to all the visual clutter that advertisements impose on us. Instead, I spend my money on sustainable products, supporting farmers and businesses that work hard to make a difference.

IT IS SO MUCH EASIER ON THE EYES

This might sound irrelevant and very superficial, but yes, a clutter-free, zero waste home just looks *so darn good*. And there is nothing wrong with that. Who wouldn't want a beautiful home?

Do you drool over those modern bathrooms, which look like little spa resorts? They always look so incredibly appealing in magazines and catalogs. Let's say you decide to spend a small fortune on remodeling your tired-looking bathroom. At the end of the renovation, it looks great. Well, that is until all these "necessary" products move in and clutter your new retreat, you know, those insanely colorful plastic bottles of shampoo, body wash, and body lotion. And for some reason, toothpaste and toothbrushes always come in bold, primary colors, ones that would clash with your new bathroom aesthetic.

Packaging is designed to catch our attention and establish a brand; it is the perfect canvas for advertisement. Each product packaging wants to stand out from the crowd, competing for our attention. Since we tend to accumulate numerous products at home, this can lead to a lot of unpleasant visual clutter.

Go easy on your eyes by going zero waste!

Zero waste helps to reduce visual clutter.

HOW TO GET STARTED

HOW TO GET STARTED

FEEL FREE TO USE THIS BOOK AS YOU WOULD A COOKBOOK. FLIP THROUGH THE PAGES TO LEARN ABOUT THE ZERO WASTE MIND-SET, FIND INSPIRATION, AND TRY A NEW RECIPE.

The most difficult part is getting started. Just start somewhere, and the rest will follow.

TAKE INVENTORY

GET TO KNOW YOURSELF!

The most important thing is to get to know yourself and your habits. This will help you diagnose the unique challenges you will face as you embark on a zero waste lifestyle and determine what is likely to be a piece of cake and what will be slightly more difficult to tackle.

It makes sense to focus on the easy fixes first, and then go from there. There really is no need to make life more complicated than it already is, right? Personally, I am a big fan of baby steps—starting out at level one and slowly working your way to the "end boss." Slow and steady wins the race! #TurtleSpeedPower!

ANALYZE YOUR TRASH

For one week, collect every piece of trash you currently create. Yes, even the trash you would have tossed outside your home. It might sound disgusting, but it does not have to be. You can either collect everything in a garbage bag or just snap a picture with your phone. Just remember to stick to one system so you can get an easy overview. Do not change any behaviors during this week. After all, this is about getting to know your daily habits.

When the week is over, it is time to inspect your trash. What have you collected a lot of? Are they wrappers from convenience food items, disposable cutlery, TV dinner trays, or to-go cups from your busy lifestyle? Record this down and take a picture as a reference for later. And please, do not feel bad about it!

Photo credit: schab/Shutterstock.com

TIPS

Additionally, you can take pictures of any items you buy, be it food or clothes. This way, you will have a record of "before" pictures that you can refer to at a later point in time. Without "before" pictures, we might be too hard on ourselves, thinking we have only achieved so little. Personally, I regret not having taken any pictures of my progress.

Consider tracking your expenses before and after you embark on your zero waste journey. This will show you exactly what you spent your money on "before" and how living zero waste will most likely save you money. You can jot it down on a piece of paper or use an Excel template.

Photo credit: Arya Photos/Shutterstock.com

Now, review your pictures and notes. The trash you have accumulated the most of are your personal "problematic areas." They are also the areas with the biggest potential! This book intends to provide simple solutions for tackling the most common problematic areas, and you can refer to sections in the book that are most relevant to you. For example, if you mostly accumulated to-go cups and other convenience products, you can jump right to the chapter "Where to Shop" (p. 43).

REDUCE, REUSE, RECYCLE

You have probably heard of the three R's "reduce, reuse, recycle." It's a great memory device! Schools, waste management facilities, NGOs, and governmental agencies all over the world use them to educate the general public about environmental sustainability.

There are many more R's out there, and many people have expanded on the three basic R's. I once even saw a very impressive and inspiring list of more than twenty R's for a more sustainable life, including *respect* and *recover*, two important aspects that receive far too little attention! Béa Johnson, the founder of the zero waste movement, also provides her own version with five R's. I want to encourage you to use these basic three R's to build your very own memory device that works for you!

For me, my personal R's are:

1. RETHINK **BE EMPOWERED!**
2. REDUCE **LESS IS SO MUCH MORE**
3. REUSE **REUSE AS MUCH AS POSSIBLE**
4. REPAIR **PROLONG THE LIFESPAN OF THINGS**
5. RECYCLE **DIVIDE AND CONQUER**

1. RETHINK

Personally, I believe zero waste is first and foremost a shift in mentality toward empowerment. In finding an open mind to try new things, we learn to challenge the status quo and to embark our own path to happiness. It is easy to dismiss the idea as too restrictive, too constraining. But I like to argue that this is a very deficit-oriented way of looking at things.

It isn't surprising that we tend to think of all the things we would have to "give up" when we first stumble upon the idea of zero waste. We are bombarded with advertisements left and right that tell us how much better our life would be if only we bought this fancy car, used that particular brand of deodorant, or drank water from that certain mountain in France. **Yet, happiness in the US peaked in the fifties.** "All in all, we have more stuff and less happiness," McKibben concludes in his book *Deep Economy.*[13]

---

13 McKibben, "Deep Economy," 35–36.

Material things will not make us happy in the long run because we get used to them very quickly and then the novelty wears off. What *does* make us happy is financial security in the sense of not having to deal with the existential threats that come with poverty. After this threshold is reached, happiness does not increase with more money.[14] However, our happiness does increase when we spend time with friends or our partner and if we have good mental health.[15] And, most interesting, doing good—giving back, lending support to others, volunteering for a cause—also makes us happy.[16] In my book, that makes a strong case for doing good, as opposed to indulging in consumerism with all its cruel externalized costs like exploitation and pollution. I know, facing a big change is scary, but living your life more in alignment with your values and discovering a whole new world along the way makes it all worth it, I promise.

2. REDUCE

We all have them at home—the bad buys, such as the neglected clothes in our closet that make us feel guilty whenever we see them, the piles of business cards with faceless names, enough pens to last us 526 years, takeout menus of places we will never order from, annoying junk mail (see p. 147), and bottles of shampoo and body wash so tiny they would look lost in a doll house.

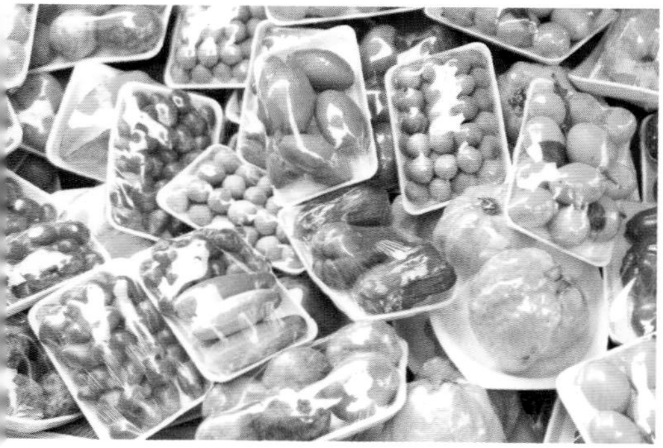

At one point, all those things had to be manufactured, packaged, and transported. And yes, all of that ate up precious resources. **Instead of hoarding unused and redundant items, does it not make more sense to give it to somebody who will put those things to good use?** By redistributing things, we do not have to use up already scarce resources to produce even more stuff to fulfill demand!

Overpackaged produce.
Photo credit: Most popular/Shutterstock.com

14 Simon-Thomas, "What is the Science of Happiness?"
15 Inman, "Happiness Depends on Health and Friends, not Money, Says New Study."
16 Simon-Thomas, "What is the Science of Happiness?"

TIPS

- Instead of accepting business cards or flyers, just take a picture of it with your phone. It will not feel like a rejection if you are friendly and say something like: "Thank you so much! You know what? Let me take a picture of it instead. This way, I always have all your information with me on my phone instead of leaving it somewhere in a drawer—and you get to reuse the card!"
- Freebies like pens or ridiculously small bottles of shampoo can be very tempting. I find it helpful to remind myself that those items are usually of very poor quality. They had to be produced at a low cost, which usually means they contain many harmful substances, were manufactured at the cost of workers, and have a dreadful environmental footprint; according to a research by the Campagin for Healthier Solutions, 81 percent of tested dollar store products contained at least one hazardous chemical above levels of concern.[17] Besides, they clutter our homes—too good to throw away, too useless to keep. Why deal with that in the first place?

3. REPAIR

We live in a day and age when gadgets are only cool until the next model is introduced only months later, when fast-fashion stores are supplied new collections every week, when buying a new printer is cheaper than replacing the ink. There is a term for that: *planned obsolescence*. Things are designed to be short-lived so we can replace them faster. But it doesn't have to be this way. A lot of items can be repaired, mended, or patched up to squeeze some more life out of them. Whenever you do make a purchase, do your homework and opt for quality and reparability.

If you do not have the skills to fix everything yourself, you can obviously have things repaired in a shop. You can also look for a so-called *repair café*, where neighbors help each other repair things! It is a great social activity, and I love how it connects people.

---

17 Taylor, "A Day Late and a Dollar Short," 3.

4. REUSE

Opt for reusables instead of single-use items!

Single-use items are great—for the companies that sell them. Items like cotton balls, wipes, or paper towels are *consumables*. This means they must constantly be replaced, and we have to keep spending money to buy them. Luckily, there is a reusable alternative for almost every single-use item!

To me, reusing things also means opting for used, pre-owned, secondhand items. **There is more than enough stuff in this world; all it takes is to redistribute them appropriately!** This way, we do not have to waste our precious resources on producing more stuff.

5. RECYCLE

After you have gotten into your new mindset (*rethink*), *reduced* your consumption by axing superfluous items and opting for quality over quantity, *repaired* what you have so you don't have to buy new items, and gotten into the habit of *reusing* as much as possible, you should be left with a lot less trash. Finally, *recycle* whatever is recyclable. Learn about your municipality's recycling policy. If they offer compost bins, awesome! If they don't, consider looking into composting at home as the eco-friendliest way to recycle unavoidable waste, such as kitchen scraps (see p. 152). Composting at home means no emissions caused by transport and no valuable resources wasted on running a big treatment facility.

FIND WHAT WORKS FOR YOU

Now it's your turn! We all have access to different infrastructures, face different challenges, and are in different places in our lives. There is no one-size-fits-all solution, and I believe being empowered is about embracing your inner badass-ness and making things your own. Here are more R's for you to choose from to make your own memory device!

Respect	Reclaim	Recover	Reevaluate
Responsibility	Repurpose	Reflect	...
Refuse	Rebuild	Reinvent	...

TAKE INVENTORY

THERE IS ALWAYS A HUNGER FOR **MORE**, BUT WHEN IS IT EVER GOING TO BE **ENOUGH**? THESE "EXERCISES" ARE AIMED TO HELP YOU ADOPT HEALTHIER CONSUMPTION HABITS.

THE PANTRY

How does your pantry look? Pretty stuffed? We have all been here: there is a recipe you just *have to* try. You go out to buy the ingredients, but because you can rarely just buy only the amount you need, you end up with a surplus of ingredients populating your pantry. Or perhaps you bought some flour, only to realize that you still have a bag of it sitting in a dark, hidden corner of your pantry. Or perhaps your love for tea has amounted to quite the collection—and there is more coming in than going out.

I CHALLENGE YOU TO USE UP AS MUCH OF YOUR STOCKPILED SUPPLIES AS YOU CAN WITHIN THIRTY DAYS.

› Ah, the power of lists! Take inventory of every item and how much of it you have. This is a good way to finally shed light into the mystery that is your pantry. Check off what you use as you go along. What a great feeling!

› Take "before" pictures of your pantry, and remember to keep your inventory list, too. These items will help you to monitor your progress. Sometimes, we tend to be too hard on ourselves, focusing our attention on the deficits, forgetting about what we have already accomplished.

Clearing out your pantry prevents food waste, saves money, and is the best way to make space for a beautiful zero waste pantry!

TIP

Use your inventory list to look up recipes online that will help you make use of ingredients you already have in your pantry, and make sure any additional ingredients can be purchased as fresh ingredients. Construct a meal plan around these fresh ingredients and incorporate them into your shopping list.

CRAMMED CLOSETS, TOO MANY PAIRS OF SHOES, AND A NAGGING FEELING OF GUILT

When it comes to clothes and shoes, it seems the sky is the limit. There is always a reason or an occasion to buy another piece. With fast fashion, clothes are short-lived disposable goods, and shopping has become both an emotional outlet and a legitimate hobby. Don't worry. You are not alone.

The truth is the majority of us have more than enough clothes. In fact, instead of buying even more fashion items, we would be a lot happier if we decluttered and downsized our wardrobes (see p. 39).

It is no secret that working conditions in the textile industry are horrendous and life-threatening. Every time we buy a piece of clothing that has not been ethically manufactured, we support those practices with our money. With our purchase, we are saying that we are okay with workers being exploited. We are saying that we do not mind the pollution it causes. We are giving them our money so they can continue to do these things.

I DARE YOU TO NOT BUY ANY CLOTHES, SHOES, OR ACCESSORIES FOR THIRTY DAYS.

Here is a neat little trick for you. If you see something you want to buy, just put it back on the shelf and walk away. If you still want to buy it after seven days, go ahead. However, oftentimes you will most likely already have forgotten about it. The urge to buy something is a fleeting sensation that vanishes just as quickly as it appeared.

OTHER SUPPLIES

If you are anything like how I used to be, clothes and food are not the only things you hoard. In our home, we had shelves and drawers full of body wash, shampoo, nail polish, cosmetics, toothbrushes, toilet paper, cleaning supplies, and cookware. And we don't even like cooking!

USE IT UP

Learn to use up what you have instead of opening a new bottle or buying something new just because you get bored of it or something new catches your eye. Once you make finishing a product your goal, you will realize that a tube of toothpaste or a bottle of all-purpose cleaner can last for an insanely long amount of time.

You might get impatient at this point if you are itching to switch to zero waste options now. That is completely understandable; once you get used to the new alternatives, you most likely will not want to go back! You always have the option of donating unused products to shelters or just giving it to family and friends.

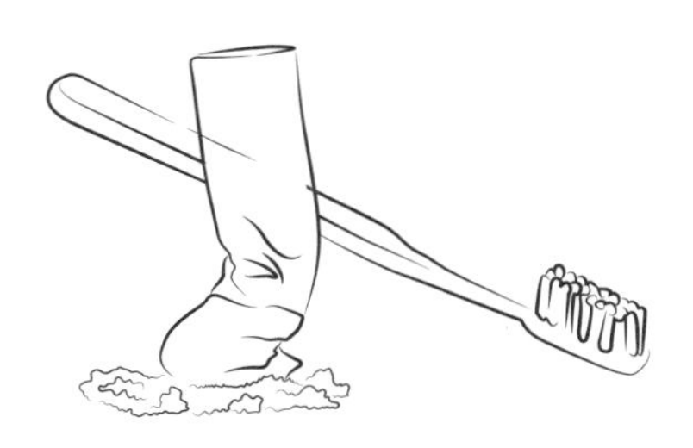

MISCELLANEOUS

My husband and I used to accumulate so much miscellaneous stuff, superfluous things nobody would miss if they were never invented. **My challenge for you: do not buy anything unless you really need it. Yes, I am basically challenging you to be lazy!**

One very effective way to do this is to avoid window shopping. We all complain about having too little time for the important things in life. Why should we waste our precious time and limited lifespan in random stores?

Another trick is to limit your exposure to advertising, be it in print, on TV, or online. Ads are basically sales pitches for primarily redundant things.

No junk mail, please

No free newspapers, either

A simple "no junk mail" sign on the mailbox can work wonders (see p. 147).

DECLUTTER YOUR HOME

Now that we have put a stop to accumulating more stuff, we can step back and reevaluate what we truly need. This step is by no means an absolute necessity if your aim is to simply reduce the amount of trash you create. However, I strongly believe that a large-scale reevaluation is the foundation for a holistically sustainable lifestyle.

Getting rid of things you own and sticking to only what you "need" can be a scary thought. I, myself, was quite a collector and perhaps a bit of a hoarder. I used to have a strong emotional attachment to my belongings. Yet, I have started to enjoy owning less and less over the last few years; I never expected this to have so many upsides! There is less to maintain. Less to be cleaned. Less to be messed up. Less to worry about. I feel so liberated and love how manageable and neat our home has become.

WHY IS DECLUTTERING A SUSTAINABLE PRACTICE?

› There is already enough "stuff" in the world. Redistributing these things means using up less of our already scarce resources to produce new "stuff."

› Bad buys, unused items, or superfluous things are just as wasteful as single-use items. Storing them at home instead of repurposing them or sending unusable things to the landfill only postpones the problem.

› Decluttering in a sensible and responsible way is not just about throwing things away; it is also about putting items that currently collect dust in your home to good use (reuse or upcycling). And if it can't be reused or upcycled, try to recycle as much of it as possible so it can replace the use of virgin resources.

You might have heard of the terms "sharing economy" or "access economy." A simple way to describe this is "access over ownership." It makes sense to *share* things that we only need from time to time (e.g. tools) or amenities (e.g. swimming pools). You can share, borrow, or swap almost everything, from sewing machines to baby clothes and kitchen appliances to sports gear on local Facebook groups, Craigslist, etc. Some communities have even set up little "libraries of things" in public spaces.

TIP

Collective acquisitions with other family members or neighbors is another great option. It strengthens the sense of community. If you happen to have spare room in your shed, garage, or basement after your decluttering project, why not make it a communal space?

ON BUSY DAYS

Grab five to ten things every day as you happen to see them, and place them in a box. Start by saying goodbye to duplicates. You can sort out the box on a quiet day.

WHEN YOU HAVE THE TIME

I personally prefer to declutter by category, e.g., shoes, office supplies, kitchen utensils, and so on. For example, gathering all office supplies in one place helps me to not forget about the ones in the junk drawer in the other room. However, many people prefer to declutter by room. Choose what items you want to keep, redistribute, and recycle.

REDISTRIBUTE RESPONSIBLY

- Set up a "redistribution" shelf in your hallway or foyer so whoever pays you a visit can take home what they might have a use for.
- Set up a little free library for things in your neighborhood, but make sure to not just dump your stuff there.
- Donate only items in good condition to social organizations (make sure these organizations do not sell these things off to developing countries, where they negatively impact local markets).
- Most public libraries accept book donations.
- Set up a garage sale or a stand at a flea market.
- Sell or give things away via local Facebook groups or Craigslist.
- There will always be items that are beyond repair or repurposing. If possible, recycle them to reduce the use of virgin resources. If not, remember that postponing their disposal does not save them from the landfill or incinerator.

WHERE AND HOW TO SHOP

WHERE AND HOW TO SHOP

MOST OF WHAT IS IN OUR TRASH CANS AT HOME TYPICALLY CONSISTS OF FOOD PACKAGING.

Learning how to avoid food packaging is a powerful way to reduce your trash. This applies not only to groceries, but also the coffee you buy on your way to work or the salad you grab on campus between classes.

"WE ARE SORRY, BUT THIS IS AGAINST HEALTH AND SAFETY REGULATIONS."

To minimize waste, we bring our own food containers out to take-out lunch spots or grocery stores, so we don't have to waste a disposable container. However, if you've ever done this before, chances are that somebody has refused your request using the excuse that it supposedly violates health and safety regulations. At first glance, it seems like there is nothing we can do; after all, we can't change regulations at a cheese counter. Frustrated, we shove our containers back into our shopping bag and never question whether or not this is actually true.

However, in most cases, it is simply not true. Referring to (sometimes nonexistent) health regulations are used by most outlets as an effective way to shut up "difficult customers." In fact, more often than not, there is no clear regulation when it comes to customers using their own containers or bags, only common practices. One health inspector may allow what another one prohibits.

Food chains, especially, set up very strict store policies because they are afraid of being sued. Bear in mind that most of the time staff turn down customers simply because they are afraid of getting in trouble or feel that it is a pain to accommodate you. Going back when somebody else is manning the counter is often an easy but effective solution!

WHERE TO SHOP

Depending on where you live, bulk options may be easy or difficult to find. Just as we all have different needs and have to face different challenges, we also have access to a different infrastructure. There is no one-size-fits-all solution. Reducing your annual trash to just a jar might simply be beyond what is possible in your situation—and that is totally fine.

Zero waste is not about perfectionism; it is about making better choices within your means. Just go for the more, or even most, sustainable option available to you as often as possible. Work toward minimizing your waste one baby step at a time.

It takes time to rediscover your area to find places where you can buy items in bulk. Once you do, you will see bulk everywhere! Start with what is relatively easy and, frankly, most essential: loose produce. Even though many grocery stores nowadays prefer to suffocate fruits and veggies in layers of plastic, there are usually loose options available.

TIP

The smaller the store, the better the chances they will accept your own container. Big-box grocery chains tend to have strict store policies, and not even the store manager has the authority to decide whether or not to accept your container.

The Zero Market, Aurora, Colorado.
Photo credit: Lyndsey Manderson

ZERO WASTE BULK STORES

Zero waste bulk stores are popping up all over North America and Europe. Every store is different, but the selection usually ranges from dry goods and household items to package-free personal care products and anything else you need for a zero waste lifestyle. If you have a zero waste bulk store in town—lucky you! However, most of us probably do not, so make sure to check out the following options.

LOCAL FOOD CO-OPERATIVES

Food co-ops are basically cooperatively owned grocery stores. Every co-op is different, but they always have a strong sense of community, they value social responsibility, and they aim to make natural foods more accessible and affordable. Some co-ops are open, i.e., everyone can shop there, but members get a discount. Other co-ops are only open to members.

Coops are community-oriented. So if you decide to join a co-op, you, too, can help shape it! Members discuss and vote on issues. If being able to buy organic and package-free goods is important to you, let others know!

Photo credit: Katja Marquard

ETHNIC (SUPER) MARKETS

It is always worth checking out your local South American, Indian, Southeast Asian, Middle Eastern, or any other ethnic grocer. Besides loose produce, they often offer dry goods like legumes, grains, rice, or spices in bulk. Some also have delicatessen counters or house-made baked goods, and they are usually very accommodating as long as you ask in a friendly and open-minded manner. Middle Eastern grocery stores are a good place to look for traditional (palm oil–free) olive oil soap (see p. 109), while you can commonly find unpackaged fresh tofu and bulk rice in Asian supermarkets.

LOCAL FARMS

You can bypass big-box stores and buy directly from local farms, supporting your local economy instead. Depending on the farm, you will be able to get produce, eggs, dairy, and meat. Depending on the season, you can also pick your own berries at U-pick farms!

FARMERS' MARKETS

Ah, I love farmers' markets! They are another great way to support your local economy (by cutting out the middleman). Farmers are usually very happy to take plastic bags or paper trays back to reuse them, since it saves them money.

HEALTH FOOD STORES

Some health food stores have bulk sections. However, their policy on customers bringing their own cloth bags and containers may vary. In my experience, the big chains are very inconsistent and seem to change their policy on this constantly. Some stores can take off the weight of your own containers at the till. If they cannot, stick to cloth bags as they are lighter. We rarely ask for permission since the worst that can happen is being told to use their plastic bags next time.

Photo credit: Katja Marquard

Did you know that . . .

. . . reducing the consumption of animal products is another great way to go easy on the planet? All animal products have a very bad carbon footprint and use up shocking amounts of resources. Did you know that, according to a UNESCO report, it takes a whopping 1,850 gallons of water to produce only one pound of beef?[15]

15 Mekkonnen and Hoekstra, "The Green, Blue and Grey Water Footprint of Farm Animals and Animal Products," 5.

BAKERIES

Bring a clean cloth bag and ask the staff to hand you the bread. You can also put your own container onto the counter and have them place the pastry in the container with tongs. Small bakeries might even sell you flour, baking powder, yeast, or seeds in bulk if you ask nicely!

TEA SHOPS

Loose-leaf teas are the obvious zero waste choice. Most loose-leaf teas can be re-steeped up to three times, and very high quality tea even up to twenty times! Yes, they are expensive, but in the end you get a lot more (flavorful) bang for your buck. Just bring your own jar and have it filled up at your local tea shop.

COFFEE ROASTERS

Just like tea shops, coffee roasters will usually be happy to fill up your own containers. Roasting coffee is a craft, and we love to show our appreciation for it. Wouldn't it be a shame to keep the coffee they handled with so much care in a layer of toxin-leaching plastic?

THE BUTCHER'S OR THE CHEESE SHOP

Small stores are almost always more accommodating than big chains. However, you still might have to compromise when it comes to meat and dairy. Even small stores can be hesitant about allowing you to put the container you brought from home onto their scale. If the store has a deli counter, you might be able to ask them to use one of their plates for the scale and let you transfer everything to your container. But of course, the eco-friendliest option is still to reduce your consumption of animal products.

Italian restaurants that make their own pasta are usually very proud of their craft. Asking to buy their pasta is a great way to show your appreciation!

ITALIAN RESTAURANTS

Depending on where you live or what grocery stores you frequent, the pasta you buy probably comes in cardboard boxes with plastic windows at best. To avoid all packaging—or if you just love the superior taste of fresh, homemade pasta—it is worth looking for an Italian restaurant that still makes their own pasta.

Making pasta from scratch yourself is another option. It is more time-consuming but also very rewarding, and you do get a good workout out of it (see p. 80).

COMMUNITY-SUPPORTED AGRICULTURE (CSA)

CSA is a model connecting producers and consumers. Consumers can subscribe to the farm, usually via a membership. They pay a monthly or yearly fee that covers the costs of running the farm, and in turn, they get their share of the harvest. This usually means a weekly or bi-weekly delivery of fresh fruits and vegetables, often hand-picked on the very same day!

This system detaches the farm from unreasonable (world) market prices that often lead to improper working conditions and shortcuts at the expense of the environment. The subscribers, in turn, get to enjoy local produce loaded with flavors and nutrients at a great value.

Many CSA farms offer family-friendly activities throughout the year, and some even offer the option of using your labor as a means to cover your subscription costs, which can be great for families that are struggling with money.

DELICATESSEN

Delis often make items from scratch and have counters where you can buy salads, chutneys, olives, antipasti, or cheese. Bring your own containers, along with a big smile, and they will most likely accommodate your zero waste needs. If you are lucky, they might even sell bulk spices and tea!

Photo credit: Katja Marquard

SPECIALTY STORES

Ice cream, chocolate, donut, candy, or cupcake shops are all great places to show up with your own container and a friendly face. Again, avoid big chains and choose to support local family-owned businesses, as they are also a lot more likely to accommodate you.

BREWERIES

Craft beer breweries often offer growler refills. They usually have machines that seal the growler with carbon dioxide to prevent the beer from going flat, so bringing a jar does not work. Trust me, I have waltzed into a brewery with a sixty-four-ounce Mason jar just to leave with my gigantic jar still empty and five extra bucks less in my pocket because they made me buy their growler instead.

DEPARTMENT STORES

This one might not always seem obvious. But some department stores have deli counters and might sell bulk candy. However, just like big-box grocery chains, department store staff may not be very flexible and could turn you down.

CHINESE HERBAL MEDICINE STORES

If you live in one of the major cities with a Chinatown, you might have seen stores with all sorts of herbs and dry goods in huge jars or bins. Don't be shy, go in and ask for the herbs you might need to brew your own root beer or your favorite herbal infusion. While you are at it, you might try to stock up on spices, too!

OIL AND VINEGAR SPECIALTY STORES

Liquids can be a tough one. Some health food stores provide refills; you might also live near an oil and vinegar specialty store. However, the selection is better described as premium, and the price tag reflects that. Organic options are scarce, too. Personally, we prefer to buy the local organic oil and vinegar options in the biggest glass bottles we can find.

FORAGING

Knowing which wild herbs, nuts, fruits, and mushrooms are edible isn't just a great survival skill. Foraging workshops are fun and can cut down on your food costs in the long run. Do make sure that foraging is allowed in that particular forest, and please do not forage in primeval forests.

DIY
Do It Yourself

DIY is a great option to keep in mind when there is no other option available. Keeping it simple is the key!

A GOOD SYSTEM TO KEEP IT MANAGEABLE

Dry Goods Store, London, UK
Photo credit: Hanno Su

› We buy fresh produce usually *once*, and occasionally *twice a week*. Just stash one or two cloth bags in your bag, your office drawer, or your car for convenience. This way, you are always ready to go!

› It is sufficient to stock up on dry goods, oil, vinegar, and coffee *once every six to eight weeks*, since these things keep very well. You might have to go to different stores to get everything, so bundling these trips and keeping the frequency to a minimum makes sense. Unlike the produce trips, you will have to plan these trips to make sure you have the appropriate amount of cloth bags, containers, and bottles for what you want to buy.

› Things like spices, tea, bamboo toothbrushes, wooden brushes, and ingredients for making your own cleaning and body care products (e.g. baking soda, washing soda, citric acid, soap, shampoo bars, essential oils) can be more difficult to come by, or perhaps you only need very little of it, so we only buy them *once a year* or even less frequently. They also take up very little space, which works well for us. Some items like bamboo toothbrushes or silk floss (see p. 117 and 122) you will most likely have to order online. Place a collective order, i.e., stock up on items and order with friends to save on packaging and transportation emissions!

This is merely a guide for you to find out what works for your situation. It takes a bit of trial and error to find your groove, but it will eventually become second nature.

ZERO WASTE HELPERS

MESH BAGS FOR LOOSE PRODUCE

- Laundry bags are very convenient because they are light and the cashier can see what's inside. However, they are made to hold socks and bras, not three pounds of carrots, so do remember to handle them with care.
- You can buy produce (mesh) bags in some health food stores or order them online from fellow zero waster Jessie Stokes at TinyYellowBungalow.com.
- We rarely need more than two produce bags at once, if any at all. If we are not buying much, we just put the loose produce into the shopping cart or basket without using any bags, and yes, we let the produce roam free on the checkout belt. Most cashiers accept it silently. We only use our bags for small things like cherries, or when we buy large quantities of potatoes or carrots.

CLOTH SHOPPING BAGS—MY SECRET WEAPON!

- I always have a clean cloth shopping bag in my personal bag. Additionally, I do not limit its use to just groceries. Though I rarely buy anything apart from food, I can stuff other types of purchases in my cloth bag.

Top-left photo credit: Katja Marquard

- A clean cloth bag also comes in handy whenever you are hungry but have forgotten to bring your own lunch or a food container. I will sometimes buy an entire loaf of bread and munch on some of it, bringing the rest home for later. Or I will buy a burrito and simply wrap it in the bag.
- We also use cloth shopping bags to buy large amounts of dry goods, such as four pounds of oats or a very bulky pound of coconut flakes.
- You can even fold your bag and use it as a plate when buying a hot dog or a waffle from a food truck. I prefer using my stainless-steel food container, but sometimes all I have on me is the cloth bag.

SMALL CLOTH BAGS OR JARS

- Small cloth bags or jars are perfect for most dry goods like nuts, oats, legumes, and even soap.
- Jars are convenient because you can simply put them into the pantry without having to transfer the goods out. However, not every store will be able to take off the weight of your own container. In those cases, you can use cloth bags, which are more lightweight.

FOOD CONTAINERS OR JARS

- Use (leak-proof) food containers or jars for *wet* items, such as olives, tofu, meat, fish, hummus, and salads.
- Use food containers or jars for *sticky* things, such as cake, pastries, cheese, sugar, and some dried fruits.
- Use food containers or jars for *powdery* things, such as flour, cocoa powder, and baking soda.

WIDE-MOUTH FUNNEL (E.G. CANNING FUNNEL)

- Trying to get flour into a jar can be messy without a funnel, so remember to bring one along.

Tip: Dishwasher salt funnels often fit jars perfectly! They are usually made from plastic, but so are most bulk bins. Plus, at least the funnel only touches the food very briefly.

OPTIONAL: A NONPERMANENT MARKER

- With a nonpermanent marker, you can write the tare weight (weight of the empty container) and PLU number (price look-up code of produce) directly onto the jar. You can also simply make a note on your phone or snap a photo for quick reference if you do not want to write it down on your jars or cloth bags or if you forget your marker.

Shop more efficiently by (permanently) writing the tare weight onto your containers and bags at home instead of having the staff tare them for you every time you shop!

OPTIONAL: A BOTTLE CARRIER FOR (SLIMMER) JARS

- We used to carry around our jars in a big bag, stuffing dish towels in between them—and sometimes a jar would fall out, yikes! A bottle carrier makes transporting jars much more convenient.
- However, depending on the bottle carrier, not all jars will fit. After all, the carrier is made for bottles and not jars.

TINS FOR TEA AND COFFEE

Tea and coffee are better kept in dark tins or containers to ensure maximum quality.

SHOPPING BAG, SHOPPING BASKET, GROCERY CART . . .

. . . (hiking) backpack, duffle bag, etc. There are so many reusable options and alternatives instead of using a disposable bag!

TIPS

- Avoid putting your produce in plastic bags while you are shopping and carry or cart loose produce to the checkout counter. This might seem unconventional, and some cashiers will not like it, but it is always an option, especially if you have forgotten to bring along your own reusable produce bags.
- Make a shopping list to avoid the temptation of buying (unhealthy) packaged foods. Avoid browsing the store.
- Wear a big smile and use genuine, kind words with the staff. We all like to be treated nicely and to feel appreciated. If you have worked in the service industry, you know how stressful this line of work is. Special requests can be tricky and may disrupt the workflow. Try to make it as pleasant for the staff as you can.
- Become a regular customer! This is so much more enjoyable for both parties. The staff will learn about your zero waste preferences, and you will not have to explain it every single time. Family-owned businesses will feel it is worth it to continue accommodating you when their service has made you a returning customer.
- It is all about the attitude! Pretending that shopping zero waste is the most natural thing in the world will actually raise *less* attention! Staff members will assume that you have already checked with a supervisor or gotten permission, or else you would not be going about it as if it were business as usual. Once, a cashier at a big supermarket chain even assumed that the laundry bag we used for produce must have been one of the chain's new additions and searched for the price label!

- Put yourself in the shoes of the staff members. Working in retail can be incredibly stressful. They have their own workflow, and special requests can disrupt that—they might not have the capacity to accommodate you at that particular moment, or they might simply be having a bad day. We try to phrase our requests in a way that offers a simple solution and, at the same time, makes saying "Sure!" easier than saying "No!" For example, we say, "Oh, I don't need that, just put it in my container, please." I try to make small talk and make them laugh. It is always more pleasant when you can laugh together, and of course people are more inclined to accommodate you when they are having a good time.
- Oftentimes the staff are simply afraid to get in trouble—and who on earth wants to risk their job? Be informed. Ask your local authorities about sanitary regulations. Being able to tell the staff at the counter that this is well within the regulations and that there is no way you would sue them for accommodating you can help persuade them.
- Avoid asking for special requests when a place is swamped. They might be able to accommodate you some other time but not when they are already struggling to get all orders out.
- We do get refused from time to time. We respect it—for now. If we feel like we were refused because the staff member did not want to get into trouble, we will most likely try again, pretending that we did not know. Sometimes your luck depends on asking the right person at the right time.

More and more businesses are encouraging their customers to bring their own containers!

FOOD CONTAINERS

BRING YOUR OWN FOOD

Food containers (3–4 cups)

- Perfect size for sandwiches, slices of cake, fruits, or an entire meal.
- When the time comes to replace your plastic food containers, consider swapping them for durable stainless steel containers. In my rather minimalist opinion, you only need one or two containers per person in your household. Stainless steel containers seem pricey at first glance, but they last a lifetime and will save you money in the long run!

Clothes and Dish Towels

- You would be surprised to learn how many simple, effective, and frankly genius ways there are to wrap things (including your head and body) in a piece of cloth! Use the tree-planting search engine *ecosia.org* to search the term "furoshiki technique," and be amazed! Since we are not as dexterous, we are happy sticking to the basics of wrapping sandwiches, burritos, and cookies in dish towels.

Jars (pint or quart)

- Our go-to choices for salads and soups.
- Wide-mouth jars are more convenient if you are like us and are too lazy to transfer your food onto a plate.

BRING YOUR OWN CONTAINER

It can take a bit of courage to ask a restaurant to put their food into your own container when ordering takeout, but unless you are at a fast food joint that does not even use real plates and cutlery, luck will most likely be on your side. If the restaurant has concerns about taking your container into their kitchen, you can ask them to serve you the food on a plate so you can transfer it into your containers yourself. This way, they are not liable!

When we are at a place for the first time, we like to bring a selection of containers and jars of different sizes so they can choose what they need. We only order in person, or else the food might already be sitting in disposable containers when we get there to pick it up.

EAT OUT

TAKE A BREATHER, YOU DESERVE IT!

Coffee to go? Takeout? Why the rush? It can be nice to take a breather and have your coffee "for here" instead of sipping it while trying to walk, spilling it all over your clothes. Go to a nice restaurant for dinner with a loved one instead of grabbing takeout on your way home. Turn your phone off. Breathe. You deserve it!

"NOTHING DISPOSABLE, PLEASE!"

In the US alone, 500 *million* plastic straws are used in the US every day![16] I pick up trash on the street every day, and guess what is the most common type of trash I find? Straws, to-go cups, napkins, and cigarette butts.

I know that this request takes courage, especially if you are a shy person. Say: "No napkin/straw, please! And could you put it in a real cup, please." Most baristas or servers will just shrug and jot down your order, but some will become visibly confused. You can say, "Just nothing disposable, please. We are trying to reduce our trash. You know, go green—yay!" This will usually make most people laugh, and they might even tell you that they think more people should be doing this or that they wish customers would order their drink "for here" if they are going to sit at a table anyway.

---

16 Parker, "Straw Wars."

Jars are great for soup.

ZERO WASTE KIT

We are all creatures of habit. Do you grab a coffee to go on your way to work or school? If so, make it a habit to take your tumbler along with you when leaving the house. Coffee tastes so much better when it does not come out of a paper cup anyway. A little goes a long way, and it is as easy as remembering your keys, wallet, or phone. In fact, try to put things like foldable shopping bags next to where you usually put your keys, or stash some in your bag or your car.

WHAT DO I NEED?

You really don't have to carry a heavy bag around to be prepared. Think of it like getting dressed in the morning. You probably choose your clothes according to the weather conditions, the occasion, and what you plan to do that day. You might put on work clothes because, well, it is a work day. Or you might put on formal wear because you are on your way to a gala. Packing zero waste is the same with those things—just make it a habit to pack what you will need that day.

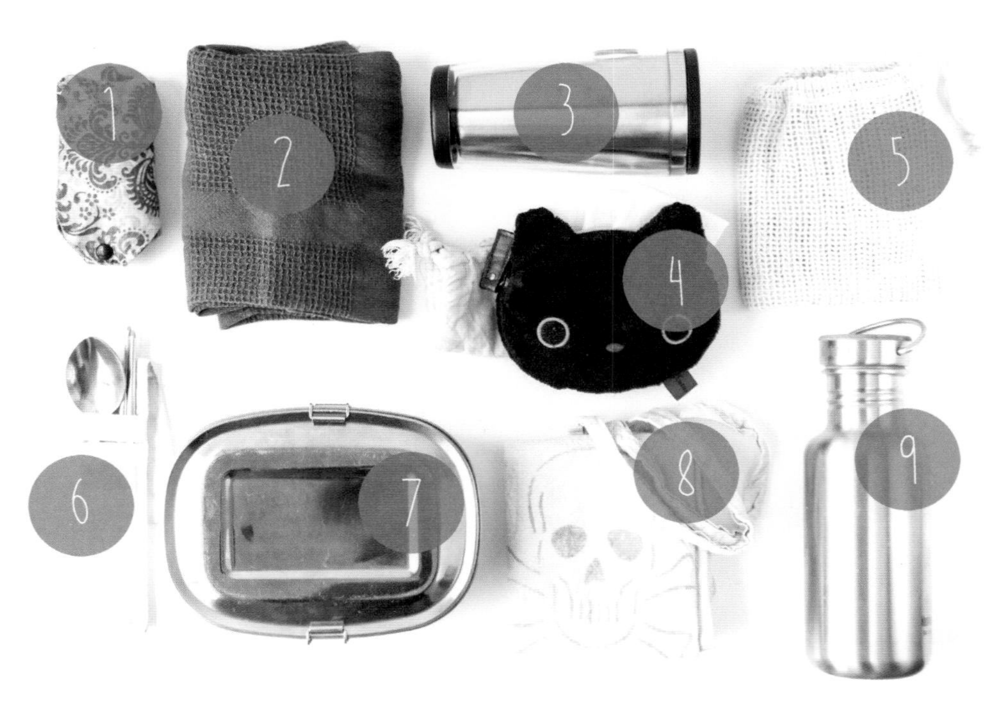

These are my suggestions for a zero waste kit. I hardly ever have all of these items with me at once—for example, I will only bring a food container if I have a lunch or dinner date or know that I might buy a pastry or sandwich. However, I will always have some handkerchiefs and a cloth bag for used hankies (point 4) in my bag.

ITEMS

1. Foldable shopping bag.

2. Dish towel. Great for burritos, sandwiches, or pastries on the go.

3. Tumbler for coffee.

4. My handkerchief purse with clean hankies that I use as napkins—to wipe my nose or to dry my hands after washing them. I also have another pouch for used hankies.

5. Mesh produce bag. You can buy mesh produce bags made from cotton or linen or just use mesh laundry bags.

6. Cutlery. You can buy travel cutlery like sporks (a spoon that you can also use as a fork) in outdoor equipment stores or just use what you already have at home. If you like Asian food, make sure to bring your own chopsticks since many places only offer disposable chopsticks.

7. Food container. Great to use when you wish to take out the food you couldn't finish at a restaurant. Of course, you can also use it for pastries or to bring your lunch to work.

8. A clean cotton shopping bag. You can use it for buying bread or dry bulk goods or simply as a shopping bag.

9. A reusable water bottle. If you are looking for a metal water bottle, go for stainless steel instead of aluminum. Aluminum bottles are a health risk. Just like a plastic water bottle, the lining of aluminum bottles can leach BPA and other chemicals into your water. What's more, the lining can be damaged when the bottle gets dented. When your water suddenly tastes funny, you are tasting the aluminum. And, no, aluminum is not considered food-safe.

FOOD PREP AND STORAGE

FOOD PREP AND STORAGE

HOW TO COOK WITH REAL, UNPROCESSED FOOD.

Going zero waste means choosing real, natural food over processed food, as that is what you will most easily be able to buy package-free.

If you are one of those avid home cooks who are into nutrition, you will love zero waste cooking.

Unfortunately, we were not. In fact, we hated cooking. We used to live off TV dinners and an insane amount of junk food. Even today, cooking feels more like a chore than something joyful. Learning how to cook from scratch was the most challenging part about our zero waste transition. However, we have found a new appreciation for food and taking good care of our bodies despite our reluctance toward preparing our own meals.

When we started out, we felt very overwhelmed. We came home with a bag full of produce and had no idea what do with all of it. At that point, neither Hanno nor I knew how to cook from scratch. We looked for recipes online, but many were too complex for us. At first, cooking was a process of trial and error, and after a while we learned that the key is to *keep it simple*!

LEARN TO PREPARE REAL FOOD—THE LAZY WAY!

Complicated recipes that require special skills are just not for us. We tend to forget about food until one of us gets hungry, and then we want something *fast*!

Instead of using convenience food to speed up the cooking process, we now rely on a few masterpieces of modern engineering (see p. 75) and, yes, on buying organic. We never expected buying only organic produce to be such a time saver! Many fruits and vegetables like carrots, cucumbers, and potatoes do not actually need to be peeled. Did you know that the outer layers often contain large concentrations of nutrients? However, if they were grown conventionally, then of course peeling produce makes sense because of the extensive use of pesticides.

We try to stick to local and seasonal food, which means we often do not buy every ingredient a recipe calls for. Instead, we get creative with whatever we happen to have on hand that week. These are some of our go-to recipes that can be prepared with whatever you can get your hands on.

FRIED RICE (APPROXIMATELY 25 MINUTES):

Let the rice cook by itself in the rice cooker (about 15–20 minutes). Chop and dice ingredients you want to put in your fried rice and fry them in a large skillet with some oil. We usually go for onions, garlic, and whatever veggies we have on hand, as well as tofu. » Put the fried veggies and tofu into a large bowl, and fry the rice in the same skillet for a few minutes. Add the fried veggies back to the pan for a final toss. » Add seasoning of your choice—some paprika, curry powder, or just soy sauce and sesame oil. » Fried rice is a perfect way to use up odd bits of veggies that are left in the fridge at the end of the week.

PAN-FRIED POTATOES WITH VEGETABLES (APPROXIMATELY 30 MINUTES):

Dice the potatoes (we prefer low-starch potatoes) and steam in a pot. Dicing them before steaming saves energy and speeds up the cooking process. » Meanwhile, chop a generous amount of onions and any other vegetables you want to add. We love to use zucchinis for this, but broccoli, peppers, mushrooms, peas, or greens work just as well. » Heat oil in a large skillet and fry the onions. Add the cooked potatoes and fry until golden brown, turning occasionally. » Remove potatoes from the skillet. Fry the rest of the vegetables in the same skillet, adding a bit of oil. Fry for 2–3 minutes, stirring constantly. » Season with salt, pepper, paprika, or fresh herbs.

PEASANT STEW (APPROXIMATELY 20-30 MINUTES):

You can pretty much make stew out of anything. Like fried rice, it is a great way to use up those odd, leftover veggie bits in the fridge at the end of the week. Cooking a stew is a very stress-free endeavor. Add ingredients that take longer to cook (e.g., potatoes, dried peas) in a pot of water or stock. While they are already simmering away, you have all the time in the world to chop the rest of the ingredients and add them as you go. It is usually ready in only 20–30 minutes. In my humble opinion, stew tastes best the next day!

CREAMY SOUPS (APPROXIMATELY 20 MINUTES):

While we love to combine our miscellaneous veggie chunks to make peasant stew or fried rice, we also love to make creamy soups whenever we buy large quantities of one vegetable, usually when it is in season and is insanely cheap. » Cut your primary ingredient (e.g., carrots) into big chunks. We like to add some nut butter, a few starchy potatoes, and sometimes lentils to add to the creamy consistency. » Cook for around 15 minutes, adding some but not too much water. » After cooking, fill up with cold water to cool down. Blend in a blender for 30–60 seconds on highest speed. Season to taste. The more powerful the blender, the creamier the consistency.

On days when we have no energy left to cook, we are content with salad, oats, bread, and olives—or just a good ol' PB&J sandwich.

Compared to processed food, we have noticed that whole foods, especially legumes and whole grain products, keep you feeling full for a longer time. In contrast, processed and convenience foods are mostly low in fiber, which will cause you to become hungry quicker. In the beginning, we started out with only small amounts of legumes and whole grain products, and gradually we added more to our diet as our digestive systems got used to it.

MEAL PLANNING AND MEAL PREPPING IS GREAT . . .

. . . but unfortunately, we suck at both—for now. This is one of the areas we are still working on. A solid meal plan can be a life-saver, especially for families—it can help you prevent food waste and save you money. It is a great way to ensure that you will use up all your groceries on time. You can streamline the cooking process by prepping all your meals during just one day of the week or by prepping the ingredients in advance whenever you have 15–20 minutes of spare time.

No stealing food on my watch!

WANT MORE CONVENIENCE? PUT YOUR APPLIANCES TO WORK!

Before we embarked on our zero waste lifestyle, we clearly did not buy heavily packaged convenience and processed food because we were craving the taste; there is no way processed food can compete with a home-cooked meal. We ate all that junk food because we were both drained and hungry after work and did not have the mental capacity left to master the complex task of cooking a meal from scratch.

It took us a while to figure out that all the soup and sauce mixes and bags of frozen precut veggies took just the same effort to prepare as fresh ingredients. And yet the taste was mediocre at best. In the end, we would have spent 30 minutes in the kitchen making food we did not enjoy.

Today, it takes us the same amount of time to cook. Even though our meals are simple, they are healthy, colorful, flavorful, and make us feel full for much longer. Instead of using convenience products to speed up the cooking process, we put our appliances to work! From a minimalist point of view, most kitchen appliances are expendable. However, if they make it more convenient for you to prepare your own food instead of choosing unhealthy, overpackaged, processed food, then they serve a purpose.

DON'T GO OVERBOARD

Before you go on a spending spree, take some time to think things through. How often would you *really* use a slicer or a spiralizer? In my opinion, one-trick ponies are worth it only if they are used multiple times a week, i.e., a toaster or a rice cooker. And chances are that you already have all you need at home; you just need to put them to good use! However, if you do decide to add another machine or tool to your hopefully very small army of appliances, buy it secondhand!

If going plastic-free is very important to you, you can find hand-crafted kitchen knives with wooden handles.

SHARP KITCHEN KNIVES PLUS A KNIFE SHARPENER

When it comes to kitchen knives, you really should prioritize quality over quantity. What good is a knife block full of wobbly, dull knives? Instead, invest in one decent, high quality chef's knife and one paring knife. Depending on the number of people in your household, you might want to increase the number to one knife per person so you can cook together.

Remember to sharpen your knives regularly!

BLENDER

Some things like almond milk or ground nuts are almost impossible to get in bulk, and others like peanut butter are often very expensive. A high-performance blender[17] can help you make those things, e.g., all kinds of plant-based milk, nut butter, turn sugar to icing sugar, and grind coffee beans and nuts. We happened to already have such a blender at home for our smoothies and shakes.

FOR COOKING

- Make creamy soups (an immersion blender gets the job done, too; however, the more powerful the blender, the creamier the soup will get, even without adding cream or coconut milk)
- Make dips and pesto (also possible with an immersion blender)
- Wet chop vegetables, e.g., cabbage (you should only do this in a high-performance blender)
- Make salad dressings and sauces

FOR BAKING

- Grind nuts (only for more powerful blenders)
- Make nut butter (only high-performance blenders)
- Turn sugar into icing sugar (only more powerful blenders)
- Grind grains to flour (only high-performance blenders)
- Make raw-vegan cakes (only more powerful blenders)

---

17 "High-performance blenders" refer to blenders like Vitamix, Blendtec, and similarly powerful blenders (which cost $500–$1,000+). "More powerful blenders" refer to decent household blenders that you can get for $100 and up.

FOR VEGANS

- Make dairy-free milk alternatives from soybeans, oats, almonds, cashews, peanuts, hazelnuts, etc. (with a more powerful high-performance blender). Soak nuts and soybeans for at least 4 hours; oats and the like do not require soaking. Toss the water. Blend 0.6 ounces of nuts, grains, or soybeans per cup of water for 30–60 seconds and strain. Always add fresh water; do not use the water you soaked them in. If you are making soy milk, you will have to cook the strained soy milk over heat for another 20 minutes.
- Make your own nut butter! Just grind peanuts, cashews, hazelnuts, and coconut flakes until creamy (do pause in between to let your blender cool off if necessary) (only high-performance blenders).

TIPS FOR MAKING YOUR OWN PLANT-BASED MILK

- You can make cottage cheese or even tofu by adding lemon juice or vinegar to soy or nut milk (this does not work with grain-based milks like oat milk).
- You can use nut butter as "instant" nut milk. Either use a whisk to mix 4 teaspoons of nut butter with 1 cup of *hot* water, or use a blender if using *cold* water. No need to strain it. Coconut butter is perfect for this method because it dissolves easily in hot water. This is a convenient way to make nut milk in small batches—making only the amount you need prevents food waste.
- If you cannot make your own nut butter and have to resort to buying them in jars from the store, it is still a great alternative instead of buying nuts or milk in cartons! One 12-ounce jar of almond butter yields 164 ounces or more than two and a half big cartons of almond milk, but it only costs slightly more than half of what the

almond milk would cost you. Jars are also eco-friendlier because cartons are very difficult to recycle. Did you know that, as the *Huffington Post* reported, almond milk can contain as little as only 2 percent of almonds?[18] The other 98 percent are mainly filtered water. Making your own nut milk from nut butter does not only taste better, but also causes less transportation emissions because no water needed to be pointlessly shipped across the continent.

- Did you know there are soy milk makers you can use to make soy and other plant-based milk, shakes, and even soups? We sold our soy milk maker many years ago when we got our more versatile high-performance blender. However, if a high-performance blender is too pricey, a soy milk maker is a good option.

MISCELLANEOUS TASKS YOUR BLENDER CAN HELP WITH

- Grinding coffee. This is not always necessary since many places that sell coffee in bulk also offer to grind it for you. However, grinding your coffee right before you brew it really gets the most out of a bean.
- "Pestling" spices. Some spices are only sold whole. If you do not have a mortar and pestle, you can also use your blender to pulverize your spices.
- Making your own (green) smoothies (with a more powerful blender or a high-performance blender) is a great alternative to buying bottled juices. Unlike juices, smoothies also make you feel full—and you still benefit from all the nutrients and fibers the fruit has to offer. However, please do not attempt to make green smoothies in a regular blender. This can shorten the lifespan of your blender or even break it!
- Shakes and frappuccinos are often only sold in disposable cups. Just make your own at home and sip it through a reusable straw. Yum!
- Pulverize baking soda, xylitol, or salt for your own homemade toothpaste or tooth powder if you are worried that these ingredients as they are might be too abrasive for sensitive teeth [page 119].

18 D'Souza, "Leading Almond Milk Brand Contains Only 2% Almonds In Recipe."

PASTA MAKER

If you eat a lot of pasta but do not have access to package- or at least plastic-free pasta, and if buying housemade pasta from your local Italian restaurant is too expensive, you might want to invest in a pasta machine. You can actually make pasta without a pasta maker, using just a rolling pin and a kitchen knife. However, this requires quite a bit of manual labor and time, and you might not be able to integrate it into your everyday life.

Most pasta makers are hand-cranked. You clamp the pasta maker to your counter and feed the dough through the roller and the pasta cutter. Operating it by yourself takes a bit of practice.

If you happen to already have a KitchenAid mixer at home, you can get a separate pasta roller/cutter attachment. Both the attachments and the packaging are plastic-free—and if you are lucky you can even find it secondhand! Having a motorized pasta maker speeds up the whole process tremendously, and making pasta from scratch will become easy and convenient.

We have a seventeen-year-old KitchenAid at home and making pasta from scratch takes us about the same amount of time as preparing store-bought ones! Though I do have to admit that store-bought pastas do not mess up the kitchen the way we do when we make pasta from scratch . . .

You can dry your homemade pasta (we use clothes hangers) and prepare them just as you would prepare store-bought dry pasta.

FOOD STORAGE

Our pantry used to be a hodgepodge of opened packages of all sizes, some tamed with clips and rubber bands, others left wide open. Plastic bags would constantly fall over, and it always took time to locate items in our crammed pantry.

For example, we would buy a new box of cereal because we could not be bothered to check how much was left in the old cardboard box in the pantry. We wasted food because we did not realize we had more than what we could finish before it went bad. According to the environmental group NRDC, the average American household of four wastes the equivalent of $2,275 of food in a year![19] Storing your food in jars can prevent that.

We do not keep our food in jars for just aesthetic or health reasons. Of course, a pantry full of jars can look neat and well organized; and yes, it is also healthier if your food is not contained in BPA-leaching plastic. But in the end, storing your food in see-through jars will allow you to easily see how much of different ingredients you still have; it also reminds you to use it up! Moreover, storing dry goods in airtight jars prevents infestations and keeps your pantry clean.

We were so used to prepackaged and hence preportioned food that we were at a loss when it came to buying in bulk. How big a jar would we need for a pound of oats? How many sixty-four-ounce jars should we buy? Or should we get gallon jars instead? Here is what works well for us. I hope it can guide you toward finding the right sizes you need, thus preventing you from purchasing the wrong jars.

---

19 Gunders, "Wasted," 12.

REPURPOSE YOUR JARS

While it is tempting to buy a beautiful assortment of mason or French canning jars, I encourage you to repurpose jars you already have. Gradually, you can thrift and collect jars that go well together and use the mismatched ones for keeping leftovers in the fridge while you phase out plastic containers. All of the jars you can see on the photo on the previous page were previously used. We asked around, saved them from our building's glass recycling box, got them off Craigslist, and purchased some in thrift stores.

It is easy to accumulate jars up to quart size. You will most likely buy food in jars because you cannot get food completely package-free. Unless you are into pickles—like, *really* into pickles—larger jars are more difficult to come by. You can try to find stores that make and sell their own pickles and ask if they have extra jars you can take off their hands. We got our sixty-four-ounce jars off Craigslist and are keeping our eyes open for some gallon jars.

Gallon jars are great to hold

- *3 pounds of flour*
- *5–6 pounds of rice*
- *Pasta*
- *Cereal*
- *2 pounds of oats*

Use sixty-four ounces jars for

- *3–4 pounds of sugar*
- *1 pound of rolled oats*

We use quart jars for

- *1.5–2 pounds of legumes*
- *0.5–1 pound of nuts*
- *1–1.5 pounds of raisins*
- *2 pounds of salt*
- *1.5 pounds of icing sugar*
- *Chocolate*
- *0.5 pounds of coffee*
- *Tea*
- *They are also the perfect size for soups and homemade almond milk*

Using repurposed jars can look less messy and cluttered if you collect ones with plain lids, i.e., lids without any prints or funny colors.

We usually use the mismatched jars for storing food in the fridge; we don't store food in plastic food containers. Sometimes we do not even bother to scrub off the labels.

Pint jars are great for storing

- *2 cups of starch*
- *2 cups of cocoa powder*
- *Baking powder*
- *Baking soda*
- *Seeds (e.g., sesame, sunflower, or pumpkin seeds)*
- *Homemade nut butter*
- *We love to drink out of both regular- and wide-mouth pint jars!*

Eight-ounce jars

- *We do not really have any use for eight-ounce jars when it comes to storing dry goods. However, we use them for homemade jams and homemade mouthwash (see p. 121), and for buying small quantities of things we just want to try.*

Four-ounce jars

- *Spices*
- *Yeast, if you like to make your own bread or ginger beer*
- *Flax seeds*
- *For homemade tooth powder, toothpaste (see p. 118), or my skin care oil mix (see p. 110)*

HOUSEKEEPING

HOUSEKEEPING
MADE EASY
WHY MAKE THINGS MORE
COMPLICATED THAN THEY
NEED TO BE?
ZERO
WASTE
WASHING SODA
BAKING SODA

You can infuse white vinegar with leftover citrus fruit peels! It takes about two weeks.

Have you ever noticed that almost all commercial cleaning products are plastered with warnings? Since I have sensitive skin, my skin would break out immediately every time I did not wear gloves when cleaning! Thank goodness for the less toxic and dangerous options out there marketed to people with sensitive skin or as eco-friendly options. But too bad that those products, too, are always sold in plastic.

Hanno and I used to stand in front of the shelves for ages, racking our brains while reading labels, trying to decipher the cryptic ingredient lists and using apps to scan the barcodes to see which product contained fewer harmful substances. It took us a while to realize that we were overcomplicating things. Instead, we started to look up traditional recipes for cleaning and reduced what we needed to only five ingredients: citric acid, white vinegar, baking soda, washing soda, and traditional olive oil soap, which is our choice for palm oil–free castile bar soap, if available.

ALL YOU NEED

These are our versatile miracle workers! They are not only great for cleaning your home but also your body!

- White vinegar
- Citric acid
- Baking soda
- Washing soda
- A palm oil free–castile bar soap (e.g., Kirk's Coco bar soap or a 100 percent olive oil soap)
- *Optional: essential oils of your choice for a nicer fragrance*

Depending on where you live, you might not be able to get these ingredients completely package- or even plastic-free. Whenever you cannot completely avoid creating trash, there is still the option of reducing the amount. Luckily, these ingredients are very yielding, and since they are also incredibly versatile, it is worth buying them in bulk.

INGREDIENT	PACKAGING	WHERE TO BUY?
White vinegar	Available in 24-ounce and sometimes 32-ounce glass bottles with plastic lids --- *In bulk: One-gallon recyclable plastic bottle*	Grocery stores, big-box stores, health food stores --- *Grocery stores, big-box stores*
Citric acid	Unfortunately it is mostly sold in small quantities in plastic bags or bottles, but it is sometimes also available in cardboard boxes --- *In bulk: big plastic or paper bags*	Canning section of grocery stores, big-box stores, health food stores, pharmacies, drugstores --- *Craft/hobby stores, baking supply stores, wine/beer making or cheese making stores*
Baking soda	Cardboard box --- *In bulk: package-free*	Baking and/or cleaning section of grocery stores, health food stores, drugstores --- *Grocery stores with bulk sections, bulk stores*
Washing soda	Cardboard box	Laundry section of grocery stores, big-box stores, supermarkets, drugstores, hardware stores --- *If you cannot get your hands on washing soda . . .*
Palm oil–free castile bar soap: coconut oil soap (e.g., Kirk's)	Recyclable paper wrapper	Your local soap maker, grocery stores, big-box stores
Palm oil–free castile bar soap: traditional olive oil soap (e.g., Alep soap)	Sometimes none at all, sometimes a paper sleeve, sometimes shrink-wrapped in plastic	Your local soap maker, Middle East grocers, health food stores
Essential oils	Small glass bottles with plastic insert and lid	Health food stores

GOOD TO KNOW	HOW MUCH SHOULD I BUY?
/ery yielding	Approximately ½ gallon per person in your household per year for cleaning purposes, provided you only use white vinegar (and not citric acid) for all recipes provided in this book
Citric acid can be hard to come by, but it might be worth the effort because it is extremely yielding! 1 teaspoon equals 5 ½ tablespoons of lemon juice or 3.4 ounces of white vinegar. It can also be used for canning, candy making, home brewing, or making your own bath bombs.	Approximately ½ to 1 cup (about 3.5 to 7 ounces) per person in your household per year for cleaning purposes, provided you only use citric acid (and not white vinegar) for all recipes in this book
Very versatile: for cleaning purposes, dental care, baking, and cooking! Quite yielding, but dwindles fast if you also use it as dishwasher detergent	Approximately 20 to 30 ounces per person in your household per year, provided you only use baking soda (and no washing soda) for all recipes in this book
Washing soda is more aggressive than baking soda, so if you substitute baking soda with washing soda in recipes (only advised for cleaning purposes!), remember to use less washing soda than you would baking soda --- *. . you can turn baking soda into washing soda by spreading baking soda on a baking dish and baking for about thirty to sixty minutes at 400°F*	Approximately 18 to 25 ounces per person in your household per year, provided you use it for all cleaning recipes in this book (including the dishwasher detergent). You can only substitute baking soda for washing soda for cleaning purposes, e.g., you will still need 5 to 10 ounces of baking soda for other uses (e.g., tooth powder)
Most bar soaps (e.g., all of Dr. Bronner's soaps) contain palm oil. Unfortunately, even organic palm oil is unsustainable. Good choices are castile soaps made from 100 percent olive oil or 100 percent coconut oil.	Approximately 20 to 30 ounces per person should be enough for both cleaning purposes and body care
Extremely yielding	Essential oils add a nice fragrance to your homemade cleaning product (an unscented cleaner can feel very odd in the beginning), but the cleaner is just as effective without it

CLEANING

DURABLE AND COMPOSTABLE HELPER

- Use cleaning rags from 100 percent cotton or bamboo fibers (you can cut up old shirts or towels)
- Dish towel for polishing
- Compostable wooden brushes
- Wooden toilet bowl brushes

Check out health food stores or sustainable/plastic-free online stores, e.g., lifewithoutplastic.com.

ALL-PURPOSE CLEANER WITH WHITE VINEGAR

According to "folk wisdom" in Germany, cleaning with vinegar can make seals and gaskets porous over time. However, I was not able to find a single study on this.

Time required:

1 minute

Ingredients:

- 5 tablespoons (or ¼ cup plus 1 tablespoon) of white vinegar
- 1¼ cups of water
- Optional: 3–5 drops of essential oil of your choice

Instructions:

Mix vinegar and water, transfer to a spray bottle, and *then* add the essential oil. Shake before use.

ALL-PURPOSE CLEANER WITH CITRIC ACID

This cleaner is odorless and does not affect rubber seals and gaskets.

Time required:

2 minutes

Ingredients:

- 1–2 tablespoons of citric acid
- 2 cups of water
- Optional: 5 drops of essential oil of your choice

Instructions:

Dissolve citric acid in water, transfer to a spray bottle, and *then* add the essential oil. Shake before use.

LEARN MORE ABOUT
HOW TO USE THE CLEANER

HOW TO USE THE ALL-PURPOSE CLEANER

Like the name "all-purpose cleaner" implies, you can use this versatile cleaner for everything in your home. Just spray and wipe (or scrub). For tough stains: spray, wait for five minutes, then sprinkle baking soda onto the surface. Baking soda reacts with the acid in the cleaner, removing stains effectively. If you want to add another scrubbing agent, use salt.

- **Kitchen:** Use the cleaner for all surfaces, including the sink, countertops, and the stove. Use a brush, baking soda, or salt for tough stains. Polish faucets with a dry dish towel.

- **Bathroom:** This cleaner removes hard water stains better than every bathroom cleaner we have ever had! And yes, "all-purpose" includes the toilet bowl. Spray thoroughly, add two tablespoons of vinegar or ½ teaspoon of citric acid directly to the water in the bowl, let sit for five minutes, scrub, and flush. You can drizzle two drops of essential oil into the toilet bowl to give your bathroom a nice fragrance.

- **Windows and mirrors:** Spray, wipe, and squeegee. If you do not have a squeegee: spray only onto stains and wipe with a wet cloth to prevent residues. Let dry and polish with a dry cotton rag (we use an old dish towel).

- **Floors:** Add ⅓ cup of all-purpose cleaner to a bucket full of water.

RECIPES

UNCLOG DRAINS

Instructions:

Mix ½ cup of white vinegar or 1 tablespoon of citric acid with ½ cup of water and set aside. » Pour four cups of boiling hot water down the drain, followed by ⅓ cup of baking soda or ¼ cup of washing soda. » Now pour the acidic mixture down the drain, cover with the lid, and let sit for five to ten minutes. » Flush with ½ gallon of boiling hot water.

TO DISINFECT AND REMOVE LIMESTONE

Time required:

1 minute

Ingredients:

- 4 teaspoons of citric acid
- 1 cup of water

Or just use full-strength vinegar instead of the citric-acid-and-water mixture.

Instructions:

Dissolve the citric acid in the water, and transfer to a spray bottle. » You can use this mixture to disinfect chopping boards or remove mildew from surfaces. » Spray, let sit for five minutes, scrub, and rinse or wipe clean.

OVEN CLEANER

Instructions:

Spray the all-purpose cleaner generously. » Let sit for five minutes, sprinkle with baking soda, and sit overnight or for at least 4 hours. » Scrub with a wet brush and wipe clean with a wet cloth.

DOING THE DISHES

DURABLE AND COMPOSTABLE HELPERS

- Use clothes made from 100 percent cotton or bamboo fibers (ideally made from old shirts or towels) instead of sponges
- Use wooden brushes with bristles made from natural materials like agave or coconut fibers instead of plastic brushes; sterilize with boiling hot water from time to time
- Use copper or stainless steel scrubbers instead of aggressive chemicals

DISH SOAP (BAR SOAP RECIPE)

This dish soap does not cut grease as well as store-bought dish soap. For greasy pans and plates, use castile soap.

Time required:

10 minutes, including grating the soap; less than 5 minutes if you have previously grated soap flakes on hand.

Ingredients:

- 1 ounce palm oil–free castile bar soap or soap flakes
- 2 cups of water
- 1 tablespoon of baking soda or 2 teaspoons of washing soda
- Optional: 2–5 drops of essential oil of your choice for fragrance

Instructions:

Grate the bar of castile soap. If you have soap flakes, skip to the next step. » Bring water to a boil, turn off heat, add grated soap to hot water, and stir with a spoon (do not use a whisk) until dissolved. » Let mixture cool to room temperature. Add baking or washing soda, essential oils (optional), and stir to combine. » Transfer to a soap dispenser. Shake before use.

DISH SOAP (LIQUID CASTILE SOAP RECIPE)

This liquid castile soap cuts grease better than the dish soap made from a castile bar. However, plastic- *and* palm oil–free liquid castile soap is even harder to come by and is considerably more expensive.

Time required:

2 minutes

Ingredients:

- ¼ cup liquid castile soap
- 1 tablespoon of baking soda or 2 teaspoons washing soda
- 2 cups of lukewarm water
- Optional: 2–5 drops of essential oil of your choice for fragrance

Instructions:

Mix to combine all ingredients. » Transfer to a soap dispenser. Shake before use.

DISHWASHER DETERGENT

Time required:

2 minutes

If you have soft water (check your local water supplier's website):

- 4 parts baking soda or 3 parts washing soda
- 1 part sea salt or dishwasher salt

If you have hard water:

- 2 parts citric acid
- 3 parts baking soda or 2 parts washing soda
- 1 part sea salt or dishwasher salt

Instructions:

Mix ingredients thoroughly to combine. » Use 1½ tablespoons per load.

You will need to use an additional rinse agent (page 98) with this diswasher detergent.

TIP

If you have a dishwasher with a dishwasher salt compartment, just fill up the compartment with salt and use 1 tablespoon of washing soda or 1½ tablespoons of baking soda per load.

RINSE AID

Time required:

2 minutes

Ingredients:

- 3½ tablespoons of citric acid
- ¾ cup of water
- 1¼ cups of rubbing alcohol, vodka, or just substitute with water

Instructions

Combine all ingredients and mix until citric acid is dissolved. Add to the rinse dispenser compartment.

TIP

You can use leftover lemon halves instead of making the rinse aid above! Just put them into the cutlery basket. The lemon halves will remove odors, but they are less effective as a rinse aid at drying out the dishes and keeping glasses from turning cloudy. You can also use undiluted white vinegar as a substitute for this mixture, but note that it is a bit less effective.

LAUNDRY

Powder detergent is often sold in big cardboard boxes. *But isn't paper recyclable?* you might think. *If so, why should I look for alternative options?* Sadly, conventional laundry detergents are quite harmful to the environment. They often contain toxic chemicals, palm oil, and nonbiodegradable additives. The big bulk boxes, in particular, contain a generous amount of anticaking agents to bulk it up. So here is how you can make your own.

POWDER DETERGENT

Time required:

7 minutes, including grating the soap; 3 minutes if you have previously grated soap on hand.

Ingredients:

- 5.6 ounces of palm oil–free castile bar soap or soap flakes
- 1 cup of washing soda
- 1 cup of baking soda
- Optional: 10 drops of essential oil of your choice for a nice fragrance.

Instructions:

Grate castile bar soap. If you have soap flakes, skip to the next step. » Combine all ingredients. Blend in blender or food processor if you prefer the soap to be almost pulverized. » Use one to two tablespoons per load.

TIP

You can use up old bar soap (e.g., hotel soap bars) for this recipe. However, do not use regular bar soap for any other recipe in this book, as they may contain additives that can lead to undesired results!

HORSE CHESTNUTS: AN ALL-NATURAL LAUNDRY DETERGENT

If you happen to live in a temperate climate zone, chances are there are horse chestnut trees in your parks! Yes, there is free, all-natural, locally grown laundry detergent literally lying on the streets around September and October! Caution: do not confuse with regular chestnuts; horse chestnuts are *not* edible.

Time required:

30–60 minutes to collect enough chestnuts for one entire year, 1½ hours to prepare the chestnuts for storage.

How much do I need to collect?

You need approximately 3 ounces of dried chestnuts or 3.5 ounces of fresh chestnuts per load. We usually wash one load per week. This is about fifty-two loads per year, which means we need 11.375 pounds of fresh chestnuts (52 x 3.5 oz = 182 oz = 11.375 lbs).

You will need:

Preparation:

- Enough horse chestnuts for one year (11.375 pounds of fresh chestnuts)
- A decent blender or average food processor (or a kitchen knife plus a large amount of superhuman patience)
- Dish towels or baking dishes

Per load:

- 3 ounces of shredded chestnuts
- Either water (option 1),

- Or a nylon sock (option 2)
- Optional: 2 tablespoons of washing soda
- Optional: 10 drops of essential oil of your choice

Instructions:

How to preserve the collected horse chestnuts:

Wash and dry the fresh horse chestnuts with a towel. » Optional: peel horse chestnuts to make sure your whites stay white. (We do not do this, because we do not have a lot of white clothes or linens. However, we have never noticed any brown stains from the peels on our white sheets.) » Shred horse chestnuts in batches in your blender or food processor (or get a good workout by chopping them with a kitchen knife). » Spread the chopped chestnuts on dish towels or baking trays. » Let dry in the sun or next to a heater. » Make sure the chestnuts are completely dry before transferring them into big jars or containers for storage, or else they will grow mold.

How to use chestnuts to wash laundry, option 1: Chestnut Tea

Put 3 ounces of shredded chestnuts into a quart jar, and fill up with hot water (you can also use cold water). » Steep for five minutes in hot water or overnight in cold water. »» Strain and use the tea as laundry detergent. » You can add up to 2 tablespoons of washing soda for heavily soiled loads. » Add up to 10 drops of essential oil of your choice if you prefer your clothes scented. » You can re-steep the chestnuts another time, but the tea will be slightly less effective. You can keep the tea in the fridge for up to a week.

How to use chestnuts to wash laundry, option 2: Bag It

Put 3 ounces of shredded chestnuts into an old nylon sock and make sure to tie an extremely tight knot. » Put the bag into the washing machine along with your laundry.

HORSE CHESTNUTS VS. SOAP NUTS

Soap Nuts are traditionally used in India. However, due to the increasing demand in Europe and North America, they have become too expensive for many locals to afford, who are forced to use aggressive chemical laundry detergents instead. The graywater often goes untreated.

Just like soap nuts, **horse chestnuts** contain saponines (**sapo** is Latin for "soap"), a soaplike chemical compound. Horse chestnuts are usually left to rot on the ground, so collecting horse chestnuts and putting them to good use is an extremely eco-friendly thing to do! It does not destroy any local markets; it does not need to be shipped across the globe; it is a great laundry detergent for sensitive skin; and it is all-natural, biodegradable, and even free—what's there not to like?

Needless to say, both horse chestnuts and soap nuts are fully compostable.

FABRIC SOFTENER

Time required:

2 minutes

Ingredients:

- 3 tablespoons of citric acid
- 3 cups of water

TIP

You can substitute undiluted white vinegar for this mixture.

Instructions:

Dissolve the citric acid in the water. » Add 3 tablespoons to the fabric softener compartment per load. » You can also use full strength white vinegar instead, 3 tablespoons per load, give or take. It is not an exact science, so feel free to adjust it to your needs.

Hand Wash Laundry Detergent

Castile soap is a great detergent for hand-washing delicate clothes. You can add one teaspoon of washing soda to the water if your clothes are heavily soiled.

TOILETRIES AND HYGIENE

TOILETRIES, BODY CARE, AND SKIN CARE

HOW ON EARTH CAN WE AVOID PLASTIC BOTTLES WHEN IT COMES TO SHAMPOO, HAND SOAP, BODY WASH, AND FACE WASH? A NOT-SO-FUN FACT: PLASTIC ONLY BREAKS DOWN INTO SMALLER AND SMALLER PIECES AFTER CENTURIES.

Bar soaps are a great way to start. You can get them completely package-free or in recyclable paper wrappers. They seem to be the perfect zero waste solution. *However*, almost all of them contain palm oil!

The demand for palm oil, the cheapest available oil, is so high that forest fires are set to illegally clear rain forests to make room for plantations, destroying natural habitats and forcefully removing indigenous people from their homes in places like Indonesia, Malaysia, and Colombia. Rain forests store more carbon dioxide than any other ecosystem in the world, so clearing rain forests releases huge amounts of greenhouse gases, contributing massively to climate change. Unfortunately, as of today, even organic and RSPO certified palm oil is far from being sustainable.

PALM OIL–FREE CASTILE BAR SOAP

Castile soap is an oil-based soap. Unfortunately, all of Dr. Bronner's soaps contain organic palm oil, and even organic palm oil is just as controversial. Your best option might be to ask your local soap maker if they make or would be open to making palm oil–free soap. Kirk's Original Coco Castile bar soap is the only palm oil–free bar soap I know of, but they do not use organic ingredients. The paper wrapper it comes in is recyclable.

We are from Europe, so we swear by traditional olive oil soap made from 100 percent olive oil, sometimes with added laurel berry oil. These soaps are made using traditional methods in Middle Eastern countries, Turkey, Greece, and France, which is why you can often find them in ethnic stores.

Natural castile bar soap is so versatile!

- Wash your entire body, including your hair (see p. 124), face, and hands with castile soap.
- Use a shaving brush with castile soap to make shaving cream!
- Make your own castile-based dish soap (see p. 96)! You can also use it directly on greasy pots and pans.
- Make your own castile-based laundry detergent (see p. 100). Use it to wash delicates or to pretreat stains.
- Use it as a natural moth repellent by placing some bars in your closet and dresser.

This is my complete body care set: olive oil soap, apple cider vinegar, and oil.

BODY CARE

THE CLEANSING PROCEDURE

Natural castile soap takes good care of your entire body. Simply apply your bar soap to your washcloth, loofah, or hands. Just rub the bar a couple of times under the water when taking a bath.

SKIN CARE

All-natural skin care products can cost a small fortune. Why break the bank when you already have everything you need in your pantry?

According to the *Telegraph*, the average woman applies 515 synthetic chemicals to her body *every day*![20] Sure, there are regulations for each product, but the exposure adds up quickly with all the products we are exposed to use in our daily beauty routines. Organic cooking oils are not only safe—you can *literally* eat them—but also inexpensive compared to drugstore cosmetics.

Think versatile! We do not need a single highly specific product for each body part. You can use cooking oil around your eyes, lips, hands, feet. Depending on which oil you use, even greasy skin can benefit from the use of oil instead of a lotion.

20 Jamieson, "Women put 515 chemicals on their face and body every day in beauty regime."

PROPERTIES OF OILS

Cold-pressed/virgin sunflower seed oil

- For all skin types
- Smells like sunflower seeds
- Good for eczema-prone skin
- Anti-inflammatory
- Rich in vitamin E
- Fast-absorbing in comparison with other oils
- Good makeup remover
- *In the kitchen: great for dressings*

Cold-pressed/virgin coconut oil

- For dry, irritated, eczema-prone skin
- Smells like coconut and summer
- Melts at around 79°F/26°C
- Relieves itchy skin of eczema
- Fast-absorbing, but only stays at the outer layer of membrane
- Not a very good makeup remover
- *In the kitchen: for baking or frying purposes up to 356°F/180°C*

Virgin olive oil

- For dry, flaky, itchy, eczema-prone skin
- Smells like my favorite salad dressing
- Anti-inflammatory
- Encourages blood flow
- Absorbs slowly
- Gets absorbed into the deeper layers of the skin
- The film it creates makes it a good lip balm and protects the skin in the winter
- Good massage oil
- Good makeup remover
- *In the kitchen: great all-rounder for both salads and frying (up to 356°F/180°C)*

Canola oil

- For dry, sensitive, irritated, flaky skin
- Has a faintly nutty smell
- Vitamin E, vitamin K, provitamin A
- Protects the skin against the damaging effects of free radicals, good antiaging properties
- *In the kitchen: great all-rounder for both salads and frying*

Sesame oil

- For dry skin with bad blood flow
- Smells like sesame
- Rich in vitamin E
- Absorbs only slowly, but it gets absorbed into the deeper layers of the skin
- Good massage oil
- *In the kitchen: great for salad dressings, for frying, or to complement a seasoning*

Soybean oil

- For normal to dry skin, combination skin, slightly greasy skin
- Almost no odor
- Soybean oil emulsifies, so applying it after a shower when the skin is still wet maximizes the moisturizing effect
- Helps regulate calluses
- Relatively fast-absorbing
- *In the kitchen: great for frying*

Walnut oil

- For combination skin—dry, flaky, but also greasy skin
- Smells like walnuts
- Very good for irritated and sensitive skin
- Rich in vitamin B
- Fast-absorbing and spreads very evenly
- *In the kitchen: great in salads*

You can mix oils to combine their properties. I like to add a bit of coconut oil to whatever I am using to add a nice scent, but you could also just add some essential oil. Tea tree oil can treat blemished skin, while mint oil cools your skin in the summer, which is an excellent relief for tired feet.

DEODORANT

This deodorant recipe is by Jasmin Schneider, author of the German DIY blog schwatzkatz.com. I am lucky to have odorless sweat, so I do not need to use any deodorant even when I am working out. Hanno, however, is not as lucky. And he swears that this is the most effective deodorant he has ever used! As somebody who sits, stands, and sleeps close to him, I can confirm this.

RECIPES

SPRAY-ON DEODORANT

Time required:
2 minutes

Ingredients:
- 1–2 teaspoons of baking soda
- ½ cup of water, sterilized by boiling and cooled down to below 120°F
- A spray bottle
- 8–10 drops of essential lime or sage oil (the deodorant is less effective without it!)
- 2 drops of tea tree oil (tea tree oil has antibacterial properties)

Instructions:
Dissolve baking soda in the water, transfer to the spray bottle, *then* add the essential oils. Shake to combine. » Shake before use.

Important: Your clothes should be free from any residues of store-bought deodorant because those residues can chemically react with your homemade deodorant! You can remove these kinds of residues by soaking your clothes in a mixture of citric acid and warm water before you put it into the washing machine.

FACIAL CARE

See p. 110 for skin care.

FACE MASK

Time required:

2 minutes

Ingredients:

- 1 tablespoons of bentonite or another kind of medicinal clay
- 1 teaspoon of water or chamomile tea

Instructions:

Mix clay and water or tea. Apply to face, neck, and chest. Let it dry, and rinse with lukewarm water.

The mask is also great to scare little kids. The mummy is back, rawwrrr!

MAKEUP REMOVER

- Use washable cotton rounds instead of disposable makeup remover wipes. Sew them out of 100 percent cotton cloth scraps or buy them from fellow zero waster Jessie Stokes's online store, TinyYellowBungalow.com. You can also find some on Etsy.
- I find that sunflower seed oil and canola oil work very well as makeup removers. While coconut oil is all the rage at the moment, it is not the best makeup remover.
- If you do not use waterproof, super long-lasting makeup, washing your face with a simple washcloth and a bit of castile soap will do the trick.

LIP BALM

Time required:

3 minutes

Ingredients:

- 1 tablespoon of coconut oil
- ½ teaspoon of olive oil
- ½ teaspoon of sunflower seed or canola oil (or just substitute with olive oil)

Instructions:

Melt coconut oil (melting point is around 78°F) and mix with olive oil and sunflower seed oil or canola oil. » Transfer to a small container and let it set for twenty-four hours.

The downside of this lip balm is the fact that it liquefies when the temperature rises above 78°F. However, it is vegan and free of both palm oil (refer to p. 109 for more on palm oil), and paraffin, i.e., mineral oil.

DENTAL HYGIENE

BAMBOO TOOTHBRUSHES

Bamboo isn't only one of the fastest-growing plants; it also has natural antibacterial properties. This makes bamboo the perfect material for toothbrushes.

However, most of the brands claiming to have 100 percent biodegradable bristles are still (perhaps unknowingly) selling bristles that contain plastics by their suppliers. You can test this yourself if you happen to have one of these toothbrushes: burn the bristles—if they smell horribly plasticky and melt into a black-ish mass, they contain plastic.

The only truly compostable toothbrush uses *pig hair*. However, pig hair is hollow inside, making it the perfect breeding ground for bacteria. Besides, we prefer a cruelty-free option. We use the bamboo toothbrush of the LA-based company Brush with Bamboo because they have been able to at least minimize the plastic content in the bristles.

If you cannot find bamboo toothbrushes in any stores near you, do ask them to consider adding those to their assortment. Let your voice be heard, and you will be able to effect change!

MISWAK OR NEEM STICKS AS 100 PERCENT NATURAL AND COMPOSTABLE OPTIONS

If you are open to exploring other options, miswak or neem chewing sticks are 100 percent natural options. You do not need any toothpaste since both neem and miswak woods contain natural compounds that protect your teeth.

You can purchase Florida-grown neem sticks sold completely plastic-free from BrushWithBamboo.com. So far, I have not been able to get my hands on miswak sticks without any plastic packaging though.

TOOTHPASTE & TOOTH POWDER

Commercial brands of toothpaste contain a wide array of chemical compounds like surfactants, preservatives, abrasives, artificial coloring, emulsifying agents, flavorings, thickeners, the hormone disruptor triclosan, and sometimes even micro plastics. You really do not need any of these for healthy teeth!

INGREDIENTS

Baking soda

- Some commercial toothpastes also contain baking soda.
- Neutralizes the acids that attack our teeth, protecting enamel from decay.
- Whitening agent, but only has a low abrasion score.
- Can easily be bought in a cardboard box.

Bentonite clay reduces acidity in the mouth, much like baking soda. It is mineral-rich and not too abrasive. However, it is sometimes difficult to get it completely out of your mouth after brushing your teeth with it.

Essential tea tree oil has antibacterial properties and helps keep gum inflammations at bay.

Essential mint oil prevents bad breath.

Xylitol is a natural sweetener and reduces cavity-causing bacteria. However, it is rarely sold plastic-free.

Coconut oil has anti-inflammatory and slightly antibacterial properties because of its lauric acid content.

Please note *that the combination of* **salty** *baking soda and* **sweet** *xylitol can take some time getting used to.*

TOOTH POWDER WITH BAKING SODA

RECIPES

Time required:

1 minute

Ingredients:

- 1 tablespoon of baking soda
- 1 teaspoon of xylitol (optional)

Instructions:

Pulverize all ingredients in a blender, and transfer to a small jar or a shaker.

How to brush your teeth with it:

Slightly dip a small corner of the wet tip of your toothbrush into the tooth powder. The powder will stick to your toothbrush, leaving the rest of the tooth powder untouched. » Brush teeth as usual. This tooth powder does not lather.

TIP

If you do not have a blender and are worried about baking soda being too abrasive, you can dissolve the tooth powder in your mouth before you start brushing.

Even though the use of fluoride is controversial, tap water is fluoridated in most areas in the USA and Canada. Check with your municipality to know if your water is fluoridated.

ANTIBACTERIAL TOOTHPASTE

Time required:

5 minutes

Ingredients:

- 1 tablespoon of baking soda
- 1 teaspoon of xylitol (optional)
- 2 tablespoons of coconut oil
- 5 drops of essential tea tree oil
- 12 drops of essential mint oil

Instructions:

Pulverize baking soda and xylitol in the blender. Slightly melt coconut oil over heat if it is too hard to stir. » Mix all ingredients together.

FLUORIDE—HARMFUL OR A NECESSITY?

The use of fluoride in tap water is a controversial topic. Fluoride can be toxic to the body and the environment. My dentist recommends fluoride in small amounts, saying, "The dose makes the poison."

Since tap water is fluoridated in most areas in the USA and Canada, using fluoridated toothpaste is not an absolute necessity. Besides, there are alternative options, such as xylitol, which serves the same purpose as fluoride to prevent cavities. You can buy xylitol in some health food stores or order xylitol for pickup in big-box grocery chains. The packaging is most likely plastic, but overall it is still a lot less wasteful.

ANTIBACTERIAL MOUTHWASH

Time Required:

1 to 2 minutes

Ingredients:

- 1 cup of water, sterilized by boiling and cooled down to below 120°F
- 1 teaspoon of baking soda
- 5 drops of essential tea tree oil
- 5 drops of essential mint oil
- 1 teaspoon of xylitol (optional)

Instructions:

Put all ingredients into a jar or small bottle, and shake. » Shake before each use. » Swish one tablespoon in your mouth for one to two minutes.

This mouthwash does not contain any preservatives! Do not make big batches, and use it up within two weeks.

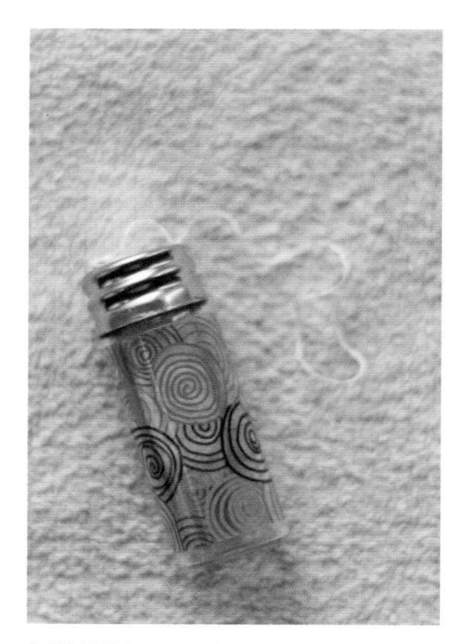

DENTAL FLOSS

- So far, there is no plastic-free vegan option for dental floss that I know of. The best you can do is go for vegan nylon floss (not biodegradable) in a cardboard instead of a plastic case (Ecodent).
- The Maine-based brand Dental Lace sells compostable, plastic-free floss made from natural silk coated with vegan candelilla wax. It comes in a beautiful glass container in a cardboard box. They also have refills in a compostable cellophane bag. You can also get natural silk floss coated with candelilla wax from Radius (sold in most health food stores). However, Radius floss comes in either a plastic case or ridiculously small sachets.
- You can unravel a piece of silk cloth, twist some threads, and use to floss your teeth.
- If your interdental space allows it, grease a tear-resistant cotton thread with coconut oil and use it as floss.

TONGUE SCRAPER

You can buy a stainless steel tongue scraper (online) or just use a tablespoon to do the scraping.

HANDKERCHIEFS

Handkerchiefs are often perceived as unsanitary and outdated, but this assumption is unfounded. Handkerchiefs are only as unsanitary as their user. Just make sure you don't sneeze into the same hanky (or tissue) for days.

HOW TO WASH HANDKERCHIEFS

Handkerchiefs are so small that we always have space to fit them into the washing machine along with our other laundry. This means they usually get washed with warm, not hot, water. As a result, we also make sure to wash them with hot water along with our whites. Washing your handkerchiefs with hot water is recommended when you have been sick.

If you don't have many white clothes or very soiled clothes that you need to wash with hot water, just put your handkerchiefs into a big bowl and pour boiling hot water over the pile. Let soak for fifteen minutes to kill off all bacteria, then wash with your other clothes.

HOW TO CARRY HANDKERCHIEFS

Just as you do with tissues, remember to have handkerchiefs handy wherever you go. We like to use a small bag or pouch for clean ones and another bag/pouch for used ones.

Where to buy handkerchiefs in this day and age?

- DIY: make them from fabric scraps or old clothes
- Ask your grandparents where they buy theirs
- Craigslist or Ebay
- Sustainable/zero waste stores or online
- Sometimes dollar and big-box stores stock them

HAIR CARE

ALTERNATIVES TO SHAMPOO IN PLASTIC BOTTLES

- Shampoo bars: run the bar over your wet hair and scalp; it should feel very similar to liquid shampoo. You can buy shampoo bars even in big-box stores.
- Castile bar soap: use like regular liquid shampoo, but make sure to apply a vinegar rinse (see p. 127).
- "No poo" options: there are several options to go “no poo.” The most popular one is using a mixture of baking soda and water. I, however, prefer using light rye flour.

HOW TO WASH YOUR HAIR WITH LIGHT RYE FLOUR

Personally, I swear by this all-natural method because I was finally able to get rid of my dandruff and greasy hair. However, it can be difficult to find bulk or even plastic-free rye flour. You can try your luck in bakeries, fingers crossed!

Warning: Hard water can cause dandruff-looking residue in the hair!

Depending on what products you used to use and your skin and hair type, it can take up to a couple of weeks to get used to the new treatment. Afterward, your scalp will have learned how to regulate itself and your hair will not get greasy as quickly as it did anymore.

When switching to this method, your hair might feel waxy. This is due to the residues from conventional hair products. Unfortunately, most of it is permanent and will not fade. The good news is that you will be able to tell that the newly grown hair is soft and very healthy.

If you have sensitive skin (like I do), then most likely both commercial shampoo and castile soap will irritate your scalp. On the flipside, rye flour does not disrupt the natural pH balance of our hair and skin and is rich in vitamin B5, which has regenerative and anti-inflammatory properties.

TIP

The rye flour mixture is also a great face mask and can be used as a body wash!

RYE FLOUR SHAMPOO

> The rye flour mixture does not lather up and is unscented. This can take some time getting used to.

Time required:

1–2 minutes

Ingredients:

- Whisk
- 1–3 tablespoons of light rye flour
- Some water

Instructions:

Using a whisk, mix rye flour with only a little bit of water. » Whisk until there are no lumps left. » Add water and whisk until the consistency is slightly runnier than shampoo.

How to use the mixture:

Wet your hair and massage the mixture onto your scalp. Leave in for 1–2 minutes and rinse thoroughly. Apply the apple cider vinegar rinse (see next page).

DRY SHAMPOO

If you have a lighter hair color, you can generously apply cornstarch to the roots with a large makeup brush or paintbrush. Brush out all the excess starch with a hairbrush. If you have a darker hair color, you might want to mix in some cocoa powder, apply it in smaller amounts, brush, and repeat.

CONDITION YOUR HAIR WITH A VINEGAR RINSE

Ingredients:

- 1 tablespoon of apple cider vinegar or lemon juice
- 1–2 cups of water

Instructions:

Add apple cider vinegar or lemon juice to a (measuring) cup and take it with you into the shower. » After you have washed your hair, fill up the cup with warm water and pour the mixture over your hair. » You might have to use double the amount for very long hair. » Leave in for 1–2 minutes and rinse thoroughly with water.

GOOD TO KNOW

- Vinegar smooths the cuticles, decreasing frizziness. Your hair will instantly feel softer and have a smooth appearance.
- Do not use white vinegar as the vinegary odor will stay in your hair!
- If you are a passionate kombucha home brewer, you will be delighted to hear that kombucha vinegar works very well, too. Please bear in mind that you will have to adjust the amount depending on how acidic your kombucha vinegar has already become.
- If you have damaged hair (e.g., from dying your hair or a perm) you can double the concentration. A vinegar rinse is particularly good for keeping the color of dyed hair.

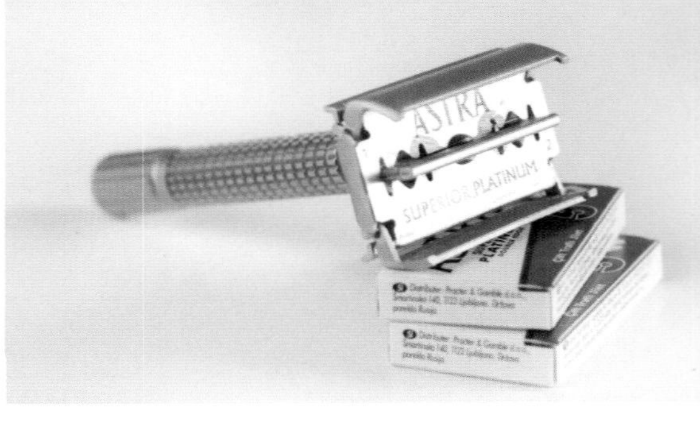

HAIR REMOVAL

The most popular method of hair removal is shaving. This usually means plastic razors and a lot of very expensive plastic razor blades. However, there are more than just this one method of taking care of hairy situations.

ELECTRIC SHAVER

Like all electric (and electronic) gadgets, the production of an electric shaver uses up a lot of resources. They also need electricity to run. But if you take good care of your shaver, it can last a lifetime.

STRAIGHT-EDGE RAZOR

This is probably the only truly zero waste option, since the blade only needs to be sharpened, and never replaced. As a traditional tool, it usually consists of materials that are either completely recyclable or compostable. However, it does require a steady hand and some fine motor skills.

SAFETY RAZOR

This is the traditional version of the consumerist plastic razor. The razor blades are 100 percent metal and some brands (e.g., Astra) sell them plastic-free, wrapped in paper only.

TIP

Always make sure to dry the blade after shaving so it does not get rusty and stays sharp. A razor with a "butterfly" mechanism will let you access the blade without any hassle.

SHAVING FOAM

There really is no need for canned shaving cream. Just use a bar of Castile soap and a shaving brush. Wet the brush and brush the soap briskly to create "true lather."

SUGARING

Sugaring or sugar waxing is a traditional epilation method. A sugar paste is made from sugar, lemon juice, and water. Getting the consistency and temperature right can take a bit of practice.

EPILATOR

An epilator can be a convenient way to get rid of unwanted hair. Like the name suggests, it is just as painful as waxing and sugaring.

PERMANENT HAIR REMOVAL

There are different methods of permanent hair removal; the most popular ones are probably IPL and laser epilation. Both methods are neither pain-free nor cheap and take multiple sessions. If you are interested make sure to thoroughly read up on it and book a consultation before jumping in.

PERIOD TALK

PERIOD TALK

THE AVERAGE WOMAN USES ABOUT 11,000 TO 17,000 TAMPONS OR MENSTRUAL PADS IN HER LIFETIME.[21]

This is not only very costly; it also consumes a great amount of resources and is potentially harmful to the body.

Photo credit: ludmilafoto/Shutterstock.com

21 Spinks, "Disposable tampons," or Mercola, "Women Beware."

HEALTH RISKS

Most commercially available tampons and pads are made from conventional cotton and cellulose. According to the Organic Trade Association, cotton is one of the world's most polluting crops due to its extensive use of pesticides.[22] Cellulose is extracted from wood, which, sadly, is often illegally sourced. The extraction process relies heavily on chemicals. This means that both tampons and pads contain harmful substances that should not have any business being so close to sensitive body parts.

TSS (toxic shock syndrome) is a bacterial infection that can be caused by the use of tampons. It is rare, but serious to a point where it can be deadly.

ENVIRONMENTAL IMPACT

Cellulose for products like paper, menstrual pads, and tampons is made from wood. WWF estimates that 15 to 30 percent of all wood traded globally comes from illegal logging.[23] Cotton is not a lot better. It is a thirsty crop, often grown in dry areas. Up to 2,640 gallons of water are necessary to produce only one pound of cotton.[24] Most jeans weigh about one to two pounds.

The raw material is then shipped across the globe in order to have it further processed, employing even more water, energy, and numerous other resources. The excessive use of chemicals pollutes the environment. The end product needs to be packaged. The packaging itself, if plastic, had been produced with massive effort, starting with the drilling of fossil fuel from the ground. But let's not get sidetracked. The final product is once again transported to warehouses and processed through a complex supply chain until it finally reaches the store, where it is transported yet again to your home. In the end, it is used for a few hours before it starts its next journey to the landfill.

22 Organic Trade Association, "Cotton and the Environment," 1-3.
23 World Wildlife Fund, "Illegal Logging."
24 The Water Footprint Network, "Product Gallery."

MENSTRUAL CUPS ARE THE BETTER VERSION OF TAMPONS

Menstrual cups are small silicone cups. Similar to tampons, they are inserted into the vagina, where they catch the blood instead of absorbing it. Menstrual cups do not irritate the vaginal mucosal tissue because they are nonabsorbent, and neither do they have any negative impact on the vaginal flora, thus preventing TSS (toxic shock syndrome), the disease that tampon boxes warn us about. Almost all menstrual cups are made from medical grade-silicone and latex. This means that, unlike tampons or pads, they do not leach any harmful substances, preventing irritations, vaginal mycosis, and allergic reactions.

A menstrual cup costs between fifteen and forty dollars and can be used for up to ten years.

As with tampons, inserting and removing the cup takes a bit of practice. Unlike a tampon, though, the cup can stay in the body for up to twelve hours, which makes it perfect for sleeping with it at night and during heavy periods. The cup is emptied into the toilet or a sink, rinsed with water or wiped with toilet paper (if you happen to be in a public bathroom), and reinserted. It is sterilized in between periods, usually by boiling it.

Photo credit: Yulia Grigoryeva/Shutterstock.com

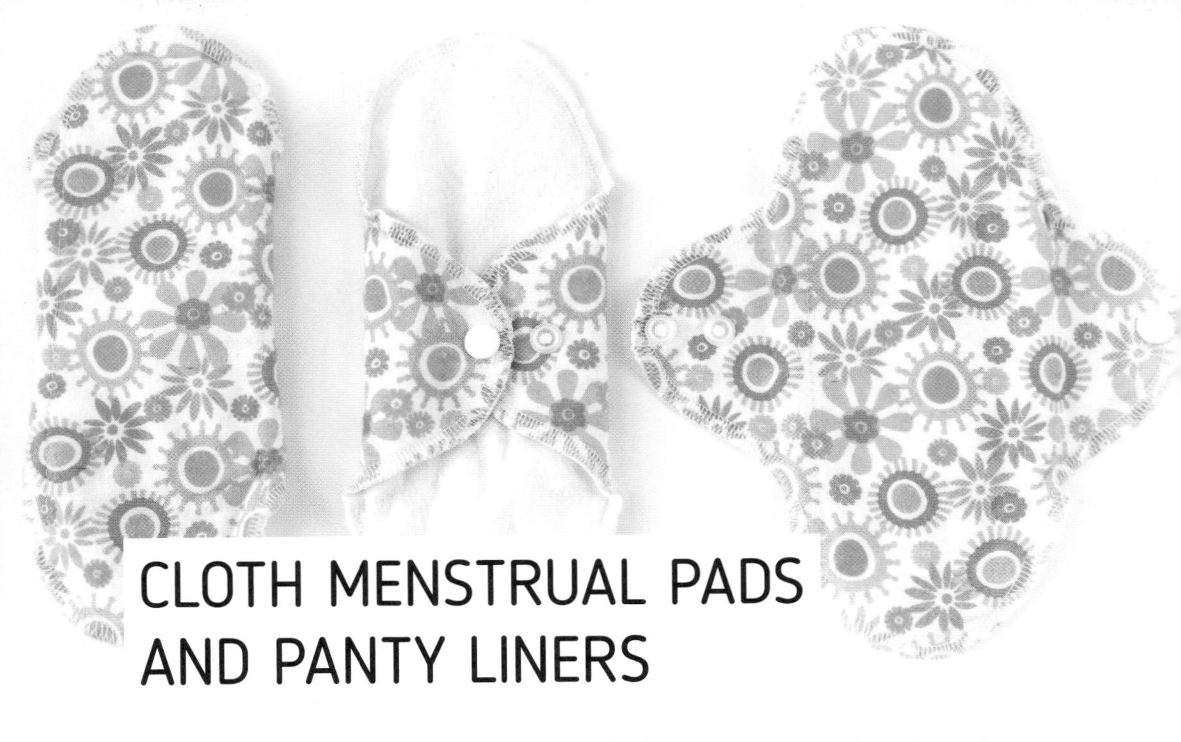

CLOTH MENSTRUAL PADS AND PANTY LINERS

Cloth pads and panty liners come in all sizes, patterns, and designs. I would rate them to be just as comfortable as their disposable counterparts.

You can buy them on Etsy and sustainable online stores (e.g., fellow zero waster Jessie Stokes's store TinyYellowBungalow.com). They are usually handmade with love, and can be pricey. However, since they can be used for up to ten years, they will actually save you money.

Making your own cloth pads only requires very basic sewing skills. Why not give it a try?

BATHROOM TALK

NUMBER ONES AND NUMBER TWOS

WE ALL DO IT, BUT WE HARDLY EVER TALK ABOUT IT.

Toilet paper is obviously a consumable and disposable product usually sold in plastic. Most people will probably not feel comfortable not using toilet paper. However, from a sanitary point of view, toilet paper is not actually the best option. But first, let me tell you where you can get plastic-free toilet paper rolls.

WHERE TO BUY PLASTIC-FREE TOILET PAPER

You can buy unbleached toilet paper made from 100 percent recycled paper, individually wrapped in paper, at office, hotel, and restaurant supply stores or online. You can probably even order some for pickup in your big-box store.

Photo credit: Andrey_Popov/Shutterstock.com

WHAT TO USE INSTEAD OF PAPER TOWELS IN A PUBLIC BATHROOM

I have two small handkerchief-towel hybrids that I purchased in Tokyo. In Japan, using a handkerchief to dry your hands is very common, so stores have started selling these hanky-towel hybrids. However, a very small face towel can also do the trick. Or, you can use a handkerchief like Hanno does.

WHY ON EARTH WOULD ANYONE NOT WANT TO USE TOILET PAPER?

Now, back to our discussion on toilet paper. Yes, you can even go zero waste in that department. And no, it does not have to be gross. In fact, it is much more sanitary than wiping (and smearing) the areas in question with toilet paper, sometimes leaving bits of soiled tissue behind. Toilet paper may contain bleach and other chemicals that—along with the mechanical wiping motion—can irritate these sensitive areas. Using lukewarm water is said to be the most hygienic and recommended method to clean yourself, as your physician will surely confirm.

Bidet bottle

I know, our nonsensical disgust toward even the thought of not using toilet paper is deeply rooted. Hanno and I probably would not even had given the idea a chance if we had not previously lived in Japan for one year, during which we came to appreciate the cleanliness of using a bidet. I mean, how often do *you* emerge from a number two feeling so clean downstairs as if your private areas had just taken a shower? (Which, in fact, they have!)

LOW TECH: THE BIDET BOTTLE

A bidet bottle, or portable (travel) bidet, is an inexpensive plastic bottle with a nozzle that you can buy online for about ten dollars. Fill the bottle with lukewarm water, point the nozzle to your backside, and squeeze. Make sure to clean thoroughly and dry with a washcloth.

THE MIDDLE GROUND: THE HANDHELD BIDET SHOWERHEAD

The bidet shower head, also sold as a "diaper sprayer," is a small showerhead that you can install on your toilet. You might be able to buy it in your local hardware store.

TIP

Search for the words "bidet sprayer," "bidet showerhead," "diaper sprayer," or "handheld bidet" to find and purchase a bidet showerhead online.

HIGH TECH: JAPANESE TOILETS AKA WASHLETS

You can choose to install a Japanese toilet in your home—a spa and wellness throne instead of just a mere toilet seat. Starting at around three hundred dollars, they come with a variety of functions, including—but not limited to—**a bidet feature!**

You can customize how you want the jet of water to clean your sensitive areas: water temperature, water pressure, angle, massage feature, male or female anatomy, front or back, etc. After the procedure, there is even a dryer option that is also customizable.

WARDROBE AND FASHION

WARDROBE AND FASHION

CLOTHES ARE USUALLY SOLD PACKAGE-FREE. WHY DO WE NEED TO TALK ABOUT CLOTHES AND FASHION, THEN?

Photo credit: Lolostock/Shutterstock.com

Clothes are commonly individually wrapped in plastic, unwrapped in the storage area of the store, and put onto hangers. But the more important reason why I choose to talk about **fashion in this book is because fashion is one of the world's largest and most polluting industries**. Fast fashion has taken the world in a heartbeat. We produce more, faster, and cheaper than ever, while refusing to pay for the true cost.

Recommended documentary: True Cost *(2015)*

Most of us know about the scandalous conditions in the fashion industry, about child labor still being the norm and not the exception. Yet, it slips our minds whenever we see those must-have shoes in the store. This is very human. After all, the suffering on the other side of the globe just isn't a part of our daily life. However, this does not mean it is any less real.

Sadly, clothing that contains harmful substances is also the norm and not the exception. Yes, children's clothing included. It makes sense when you think about it. Most pieces of clothing nowadays are made from synthetic fibers. And often this means it is no more than woven plastic. Naturally, all the yucky and unhealthy stuff you find in plastic can also be found in clothing: BPA, phthalates, flame retardants, etc. Conventional cotton is just as bad, seeing that it is the crop treated with the most pesticides on this planet. And it is a thirsty crop on top of that: on a global average, 1,200 gallons of water are necessary to produce only one pound of cotton, and up to more than 2,640 gallons in India, according research done by the Water Footprint Network.[25] Yet cotton is often grown in dry areas.

FIND YOUR BALANCE

For some reason, a crammed closet has become part of our modern culture. Before I embarked on my zero waste lifestyle, I considered myself to have a rather small wardrobe, and yet I easily reduced it by 80 percent! In this culture, we are taught to believe there is always room for more. And you might have already guessed what I am getting at: *less is more.*

---

25 The Water Footprint Network, "Product Gallery."

You can find a vast number of tutorials for capsule wardrobes on the good ol' interwebs. I will not go into details here, because I feel that your wardrobe really depends on your personal style and preferences. I for one have ditched all formal wear, dresses, skirts, and high heels, but have kept my Batman tees and Space Invader socks, which might not exactly reflect your personal preference.

DECLUTTER RESPONSIBLY

You can always sell your clothes at garage sales, on Craigslist, or on Ebay. Clothing swap parties are another very fun option. Local Facebook groups are a great place to give away or swap things, while supporting your local community.

If you decide to donate your clothes, make sure the organization you donate them to does not sell any of the clothes off to developing countries, where they end up destroying local textile markets. Some developing countries have had to resort to banning the import of those clothes in order to protect their local industries. Call the organization you have in mind or ask the volunteers in the thrift store beforehand.

Calling beforehand also goes for donating to shelters. They do not have a lot of space to store donations, and oftentimes what they need most is not clothing but household and personal care items.

BUY SUSTAINABLY

It is sad to think about, but essentially all the clothes you can buy in your local store or at the mall are unsustainable and potentially harmful. Buying secondhand is a great way to prevent the resources that already went into producing the clothes from going to waste. Secondhand clothes also contain a lot less harmful chemicals because they have already been through a couple of washing cycles. Buying ethically produced, fairly traded, organic clothing helps support businesses that are trying to change the industry by establishing a better norm.

PAPER WASTE

REDUCE YOUR PAPER WASTE

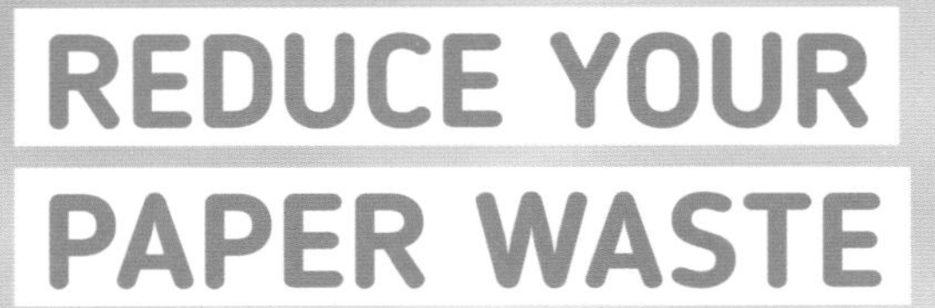

YES, IT IS TRUE THAT PAPER WASTE IS "NOT AS BAD" AS PLASTIC WASTE, BUT I AM AFRAID THIS DOES NOT MEAN IT IS "GOOD" FOR THE ENVIRONMENT.

Photo credit:
Brian A Jackson/Shutterstock.com

To produce paper, trees need to be cut down. An estimated 15 to 30 percent of wood traded worldwide comes from *illegal* logging.[26] The wood then needs to be processed with chemicals and a large amount of water. Did you know that it takes a bucket of water to produce a single sheet of paper?[27]

At the end of its lifespan, paper can indeed be recycled, usually five to seven times over.[28] However, "recyclable" and actually "being recycled" are very different things. Let's have a look at paper plates and pizza boxes. After they have served their purpose, they are too soiled to be recycled. They can still be composted, but disposable tableware is almost always thrown into the trash. We use disposable items for the purpose of convenience, and how convenient is it to then make sure you compost them? Not so much!

"NO JUNK MAIL, PLEASE!"

If you live in Canada, reducing the junk mail you receive by about 80 percent can be as easy as putting up a "No junk mail, please!" sign on your mailbox. It is usually more complicated in the US, since this is not regulated by law as it is in countries like Germany, where you can even sue a company that distributes junk mail to you when you have posted a refusal sign on the mailbox. Here is what you can do to cut down on junk mail:

- You should still slap a "No junk mail" sticker on your mailbox. It is an official "return to sender," but not every letter carrier respects that.
- Remove your name from national mailing lists at the DMA's (Direct Marketing Association) consumer website, DMAchoice.org.
- The nonprofit organization CatalogChoice.org helps you to get rid of junk mail.
- If you do not want to receive credit card and insurance offers, you can opt out at OptOutPrescreen.com.
- Opt out of receiving unsolicited Yellow Pages phone books at YellowPagesOptOut.com.
- The app PaperKarma stops junk mail for you.

---

26 World Wildlife Fund, "Illegal Logging."
27 Rep, "From forest to paper, the story of our water footprint," 12.
28 GD Environmental, "How Many Times Can You Recycle A..."

Calling the sender to file a complaint whenever junk mail finds its way into your mailbox can further your success rate. Being nice and thanking your letter carriers also help.

However, even after all those opt outs, I am afraid you will still receive addressed junk mail. Like with most things, an ounce of prevention beats a pound of cure. Only hand out your personal information when absolutely necessary. In-store reward cards, sweepstakes, and product warranty cards are usually means to collect your personal information so that those businesses can send you junk mail.

MAGAZINES

Magazines are one of those things that we cut out of our lives without missing anything. However, if you need your magazine fix, Readly is a service similar to Spotify or Netflix that lets you read thousands of magazines online for a flat monthly fee.

NEWSPAPERS

Nothing is older than yesterday's newspaper. Try to visualize all the paper wasted every day that the news is printed on! And it is not only paper—did you know that newspaper piles are often wrapped in plastic?

- Most newspapers have their own online apps, and digital-only subscriptions are often cheaper than traditional subscription options.
- Reduce paper waste by sharing a subscription with a family member or your neighbor. Offices usually get the newspaper, so why not read it at work?
- Personally, I prefer receiving my daily dose of news on my phone; I feel it is just so much more convenient.

Photo credit: Daniel Voelsen

BILLS AND BANK STATEMENTS

Nowadays, electronic bills and bank and credit card statements are very common. Mobile banking is also more convenient than ever. Take advantage of that.

PRINTOUTS

Use 100 percent recycled paper.

Unfortunately, hardly any office, school, or university uses 100 percent recycled paper. You can buy it at office supply stores for your own personal use. If you are a student, suggest the use of recycled paper to your school or university. If you are an office worker, try to convince whoever is responsible for office supplies to switch to recycled paper. However, chances are usually slim. If you run your own business, congratulations! You have all the power. May the force be with you!

Digitalize!

The day and age of keeping everything on paper is fading (at least it is in my humble opinion, as a digital nomad). Nowadays you can scan documents with your smartphone and save it directly in the cloud for synchronized access. I for one prefer looking for documents online with the help of the search bar as opposed to rummaging through physical folders.

Print two pages on one piece of paper.

Print two pages on one page and remember to print on both sides of the paper. This way you get four times the information on the same sheet of paper!

Reuse before you recycle.

I see many people taking notes on one side of a piece of paper, leaving the other side blank. Simply use both sides of the paper and cut down your tree consumption by 50 percent! That wasn't too hard, right?

As much as we like to eliminate all postal mail we get, some things still get mailed to us on paper. Almost all of those letters are only printed on one side. We scan those letters to keep a digital version for future reference, and use the back side of the paper copy for notes. We also collect the envelopes, clip them together, and use the stack as a notepad.

If you have access to a more or less "public" copy machine, e.g., as an office worker or student, you will most likely never run out of scribbling paper—there is usually a bin full of misprints next to the copy machine. Help yourself to it. It isn't pretty, but it is sufficient for jotting down a quick note or a daily to-do list.

After we have reused the paper thoroughly, we recycle it.

TRASH TALK

TRASH FACTS

- **"Recyclable" does not mean it gets recycled:** Plastic recycling is a complicated matter. Whether or not a recyclable piece of plastic gets recycled depends on multiple factors: Is there a demand for that material at that particular moment? Is it attached to another material or even just another kind of plastic? Is there a sticker on it? How small is the piece?
- **Receipts** are commonly BPA-coated, which makes them a health risk. They should not be recycled because the high concentrations of BPA on receipts will contaminate the water, soil, and whatever is made from the recycled material.
- **Beverage cartons** are marketed as 100 percent recyclable. But the truth is that they are notoriously difficult to be recycled—which is why, more often than not, they do not get recycled. Those cartons are made up of up to nine different layers glued together. The layers are so difficult to separate that there are special facilities that only recycle cartons.
- **Compostable bioplastic bags** are usually only compostable in commercial composting facilities, but not in your backyard composter.
- **Window glass, glass from picture frames, and eyeglasses** have a different melting points and do not belong in the glass recycling. They can ruin an entire batch.
- **Waste management is a logistically complex and costly system.** Waste management is an industry that does not run on the power of love for the planet. In the end, it is a business, one that requires highly specialized vehicles, bins, dumpsters, and facilities. The production of these things alone already uses up a myriad of resources. The vehicles need to be fueled in order to collect trash and to transport recyclable material across the continent. The facilities need to be staffed, the machinery powered. All of this because we seem to be obsessed with turning valuable resources into a problem.
- **Recycling is not circular!** Recycling requires a lot of energy, water, and oftentimes questionable chemicals that cause pollution. Lindsay Miles, fellow zero waster and author of the blog TreadingMyOwnPath.com, captured the essence of this when she said, "Recycling is a great place to start, but a bad place to stop."

MAKE YOUR TRASH CAN LINER OUT OF PAPER ORIGAMI

Hardcore zero wasters might not need liners anymore, because, well, you do not need those if you don't have any trash cans at home, right? But not everyone needs to join the all-in zero waste club. Besides, they can also come in handy for keeping organic waste if you are not a big fan of simply rinsing out your organic waste bucket.

COMPOSTING

If your city or town offers curbside collection of compostables, lucky you! If your city accepts your organic waste but you need to bring it to a collection point, that is less convenient but still doable. In this case, you might want to consider freezing your organic waste for convenience. Composting at home and using the compost for your backyard, however, is always the most sustainable option because it causes zero transportation emissions! Turning kitchen scraps into compost is often called "upcycling" because it turns waste into something of more value than what it had before.

If you have a backyard, a simple **compost pile** might do the trick. Apart from kitchen scraps, yard trimming can also go onto the compost pile, saving you the hassle of keeping the trimmings in your garage or shed and having to remember to put it out on collection day.

However, if you live in an apartment (like we do), you can still compost. Heck, you don't even need a balcony! We live in a small downtown apartment, and as I like to say it: yes, we got worms! A whole bin full of worms to be precise. Our **worm bin** resides in the kitchen and is home to thousands of our wiggly friends. They live in the bin, eat our kitchen scraps, hair, and nail clippings, and turn them into "worm casting," a high quality fertilizer.

GOOD TO KNOW

Unlike the regular kitchen bin, the composting process in the worm bin is odorless so you do not have to worry about any bad smells. There is also no reason to worry about worms leaving the bin. The bin is a worm's paradise, so unless there is something fundamentally wrong with your worm bin, your little friends will have no ambitions to flee their cozy home.

Cooked food, meat, dairy, onion, banana peels, and citrus peels cannot go in a worm bin. Instead, a **bokashi bin** can take care of what cannot go in a worm bin. It ferments organic material, which can then be fed to the worms or go in your backyard.

KITCHEN	ZERO WASTE OPTIONS
Aluminum foil and plastic wrap (to cover food)	Plate, bowl, tea towel, or beeswax wraps
Aluminum foil and plastic wrap (to transport food in)	Reusable food container or tea towel to wrap burritos or sandwiches in
Wax/baking paper	Grease pan and lightly dust with flour or use reusable nonstick baking mats
To-go cups	Jar for cold beverages or tumbler for hot beverages (or just a jar in a sock)
Freezer bags	Freeze food in screw-top jars, reusable food containers, or silicone freezer bags
Beverages in cartons, cans, or plastic bottles	Drink more tap water, make smoothies instead of store-bought juice, or make your own lemonade instead of buying unhealthy soda
Buying beverages from vending machines or in convenience stores	Bring your own water bottle and refill it at a drinking fountain
Coffee filters, Keurig pods	Reusable filters for drip coffee maker or French press coffee/tea maker refillable stainless steel pod
Paper liners for muffins and cupcakes	Grease and dust your muffin pan or small tea cups with flour, or use reusable silicone liners
Plastic bags for produce	Mesh bags, e.g., laundry bags, produce cloth bags
Paper napkins	Cloth napkins or handkerchiefs
Packaged bread	Clean shopping bag for buying bread in a bakery
Skewers	Reusable stainless steel skewers
Plastic straws	Reusable glass, stainless steel, or bamboo straws
Disposable food containers	Bring your own reusable food containers to your favorite joints
Tea bags, single-use tea filters	Tea strainers, French press coffee/tea maker
Vegetable peeler	Organic produce and a wooden veggie brush with agave or coconut fibers
Food storage items	Big canning or pickle jars

HOUSEKEEPING	ZERO WASTE OPTIONS
Cleaning products	All-purpose cleaner (recipe p. 91)
Wipes (for cleaning)	A rag and some all-purpose cleaner
Wipes (for body care)	A wet washcloth
Paper towels, microfiber cloths	Rags from 100 percent cotton or bamboo viscose (you can cut up old towels or sew them out of old shirts)
Trash can liners	Origami your own liners from newspapers (see p. 153) or just rinse the trash can
Plastic cleaning brushes	Wooden cleaning brushes with natural agave or coco-nut fibers
Sponges	Rags from 100 percent natural fibers, wooden cleaning brushes
Surface polish, stainless steel cleaner	Stainless steel/copper scrubber, wooden pot cleaner brush with oconut fibers
Detergent	Homemade detergent (recipe p. 100), horse chestnuts (tutorial p. 101)
Fabric softener	Homemade fabric softener (recipe p. 104)
Plastic toilet bowl brush	Wooden toilet bowl brush with agave fibers
Dish soap	Homemade dish soap (recipe p. 95)
Dishwasher detergent	Homemade dishwasher detergent (recipe p. 96)
Plastic broom	There is nothing wrong with continuing to use your old broom, however if you are looking for a plastic-free option, look for a straw broom or make one out of twigs
Dust pan	There is nothing wrong with continuing to use your dust pan, however, if you are looking for a plastic-free option, there are metal dust pans

BATHROOM	ZERO WASTE OPTIONS
Deodorant	Homemade deodorant (recipe p. 114)
Body wash	Palm oil–free castile bar soap
Exfoliating body or face scrub	Luffa sponge, coffee grounds
Hand soap	Palm oil–free castile bar soap
Face cleanser	Palm oil–free castile bar soap
Shampoo	Shampoo bar, palm oil–free castile bar soap or light rye flour (recipe p.126) plus vinegar rinse (recipe p. 127)
Hair conditioner	Vinegar rinse (recipe p. 127)
Lotions, moisturizers	Cooking oils (see p. 111)
Chapsticks	Homemade lip balm (recipe p. 116)
Makeup remover	Cooking oils (see p. 111)
Makeup remover wipes	Washable cotton rounds and cooking oil
Nail brush	Wooden brush with natural fibers
Plastic razor	Electric razor, old-school safety razor with razor blades in paper wrappers, permanent hair removal, etc. (see p. 128)
Shaving foam	Palm oil–free castile bar soap plus shaving brush
Daily or monthly disposable contact lenses	Eyeglasses, hard contact lenses, laser operation for eyes
Menstrual pads, panty liners	Cloth pads, cloth panty liners
Tampons	Menstrual cup
Tissues	Handkerchiefs
Toilet paper, wet toilet paper wipes	Bidet/washlet and washcloth (see p. 140)
Q-tips	Putting things in your ear is bad for you! If you still insist, use a bamboo ear spoon or a metal earwax extractor
Plastic toothbrush	Bamboo toothbrush, miswak, or neem stick (see p. 117)
Toothpaste	Homemade toothpaste or tooth powder (recipes p. 119, 120)
Mouthwash	Homemade mouthwash (recipe p. 121)
Dental floss	Vegan (but not plastic-free) floss, compostable silk floss (see p. 122)

OFFICE SUPPLY	ZERO WASTE OPTIONS
Writing letters	Call or email instead
Copy paper	100 percent recycled paper
Envelopes	Reuse envelopes, make envelopes from paper scraps for personal letters or cards
Mechanical pencil, regular pencils with a layer of paint on the outside	Raw wood pencils plus pencil extender
Felt-tip pens	Raw wood colored pencils
Ballpoint pen	Fountain pen with refillable cartidge and ink
Highlighter	Raw wood highlighter pencils
Notepad	Reusing envelopes and the back side of paper sheets (see p. 150)
Package tape	Natural jute binder twine
Eraser	Erasers made from natural rubber
Binders	Binders made from recycled paper
Stapler	Paper clips

ABOUT
BIBLIOGRAPHY
ACKNOWLEDGMENTS

ABOUT

Blogger Shia Su started her zero waste journey in September 2014. Lacking self-discipline and being a fun-oriented hot mess, she only ever intended to be "a tad more conscious" when it came to reducing waste. Little did she expect that it could be so easy, doable, and a shameless amount of fun! As a very vocal person, she simply could not shut up about her lifestyle, which led to her sharing her thoughts and tips on her blog *Wasteland Rebel*.

Learn more about zero waste and connect with Shia on WastelandRebel.com. For a daily dose of zero waste inspiration and a glimpse into the everyday life of a vegan, palm-oil-free and plastic-free zero waste minimalist, follow @_wastelandrebel_ on Instagram.

Photo credit: Hanno Su

MORE ZERO WASTE INSPIRATION

Zero waste should not be about whether or not you can fit your trash into a jar, which in my opinion is exceedingly overrated. It should be about choosing the more or perhaps even the most sustainable option as often as possible. It is about making better choices and cultivating more sustainable habits; it is about kindness toward others and yourself.

Join the conversation and the great community! Here are just some of my favorite zero waste content creators:

- Kathryn Kellogg *goingzerowaste.com*
- Lindsay Miles *treadingmyownpath.com*
- Christine Liu *snapshotsofsimplicity.com*
- Erin Rhodes *therogueginger.com*
- Eva Pollard *thekindplanet.com*
- Manuela Baron *"The Girl Gone Green" on Youtube*
- Ariana Roberts *paris-to-go.com*
- Anne-Marie Bonneau *zerowastechef.com*
- Gittemary Johansen *gittemary.com and "gittemary" on Youtube*
- Imogen Lucas *"sustainably vegan" on Youtube*

No zero waste list is complete without these two fabulous first generation zero waste ladies!

- Béa Johnson *zerowastehome.com*
- Lauren Singer *trashisfortossers.com*

Check out the Zero Waste Bloggers Network (*zerowastebloggersnetwork.com*) for the most complete list of zero waste bloggers around the globe. If you want to volunteer, consider joining the grassroots nonprofit organization bezero.org! Share the love and remember to support all the wonderful bloggers who put so much effort into sharing their experience, wisdom, DIY recipes, and sometimes gut-bursting funny stories!

BIBLIOGRAPHY

The Canadian Press. "Most Canadians have BPA in urine, lead traces in blood." CBC. April 17, 2013. http://www.cbc.ca/news/canada/hamilton/news/most-canadians-have-bpa-in-urine-lead-traces-in-blood-1.1303235.

Carwile, Jenny L., Henry T. Luu, Laura S. Bassett, Daniel A. Driscoll, Caterina Yuan, Jennifer Y. Chang, Xiaoyun Ye, Antonia M. Calafat, and Karin B. Michels. "Polycarbonate Bottle Use and Urinary Bisphenol A Concentrations." *Environ Health Perspect*. 117, no. 9 (September 2009): 1368–1372. https://www.ncbi.nlm.nih.gov/pmc/articles/PMC2737011/.

Congressional Budget Office. "Health Care." Accessed September 15, 2017. https://www.cbo.gov/topics/health-care.

Consumer Reports. "What to Do when There Are Too Many Product Choices on the Store Shelves?" *Consumer Reports*. January 2014. https://www.consumerreports.org/cro/magazine/2014/03/too-many-product-choices-in-supermarkets/index.htm.

D'Souza, Joy. "Leading Almond Milk Brand Contains Only 2% Almonds In Recipe." *Huffington Post*. July 2015. http://www.huffingtonpost.ca/2015/07/29/almond-milk-only-2-almond_n_7897086.html.

European Environment Agency. "Municipal waste management across European countries." EEA. November 2016. https://www.eea.europa.eu/themes/waste/municipal-waste/municipal-waste-management-across-european-countries.

GD Environmental. "How Many Times Can You Recycle A..." Accessed September 18, 2017. http://www.gd-environmental.co.uk/blog/how-many-times-can-you-recycle/.

Gunders, Dana. "Wasted: How America Is Losing Up to 40 Percent of Its Food from Farm to Fork to Landfill." NRDC Issue Paper. August 2012. https://www.nrdc.org/sites/default/files/wasted-food-IP.pdf.

Inman, Phillip. "Happiness Depends on Health and Friends, not Money, Says New Study." *The Guardian*. December 2016. https://www.theguardian.com/society/2016/dec/12/happiness-depends-on-health-and-friends-not-money-says-new-study.

Jamieson, Alastair. "Women put 515 chemicals on their face and body every day in beauty regime." *The Telegraph*. November 2009. http://www.telegraph.co.uk/news/health/news/6603483/Women-put-515-chemicals-on-their-face-and-body-every-day-in-beauty-regime.html.

Luckerson, Victor. "Here's Exactly Why Watching TV Has Gotten So Annoying." *Time*. May 2014. http://time.com/96303/tv-commercials-increasing/.

McKibben, Bill. *Deep Economy*. Oxford: Oneworld, 2007.

Mekkonnen, M. M., A. Y. Hoekstra. "The Green, Blue and Grey Water Footprint of Farm Animals and Animal Products." Value of Water Research Report Series No. 48, UNESCO-IHE. Delft, the Netherland. 2010. http://wfn.project-platforms.com/Reports/Report-48-WaterFootprint-AnimalProducts-Vol1.pdf.

Mercola, Joseph. "Women Beware: Most Feminine Hygiene Products Contain Toxic Ingredients." *Huffington Post*. June 2013. http://www.huffingtonpost.com/dr-mercola/feminine-hygiene-products_b_3359581.html.

Norddeutscher Rundfunk. "Neuland: Zu viel ist nicht genug." (TV documentation) NDR Fernsehen. 2014.

Organic Trade Association. "Cotton and the Environment." Organic Trade Association Fact Sheet. April 2017. https://ota.com/sites/default/files/indexed_files/CottonandtheEnvironment.pdf.

Parker, Laura. "Straw Wars: The Fight to Rid the Oceans of Discarded Plastic." *National Geographic*. April 2017. http://news.nationalgeographic.com/2017/04/plastic-straws-ocean-trash-environment/.

Rep, Jesse. "From forest to paper, the story of our water footprint." A case study for the UPM Nordland Papier mill. August 2011. http://waterfootprint.org/media/downloads/UPM-2011.pdf.

Schumacher, E. F. *Small is Beautiful: A Study of Economics as if People Mattered*. New York et al.: Harper Colophon Books, 1975.

Simon-Thomas, Emilia. "What is the Science of Happiness?" Berkeley Wellness by the University of California, November 2015. http://www.berkeleywellness.com/healthy-mind/mind-body/article/what-science-happiness.

Smillie, Susan. "From sea to plate: how plastic got into our fish." *The Guardian*. February 2017. https://www.theguardian.com/lifeandstyle/2017/feb/14/sea-to-plate-plastic-got-into-fish.

Spinks, Rosie. "Disposable tampons aren't sustainable, but do women want to talk about it?" *The Guardian*. April 2015. https://www.theguardian.com/sustainable-business/2015/apr/27/disposable-tampons-arent-sustainable-but-do-women-want-to-talk-about-it.

Taylor, Steve. "A Day Late and a Dollar Short: Discount Retailers Are Falling Behind on Safer Chemicals." *Campaign for Healthier Solutions*. February 2015. http://ej4all.org/assets/media/documents/Report_ADayLateAndADollarShort.pdf.

United States Environmental Protection Agency. "Advancing Sustainable Materials Management: 2014 Fact Sheet." *EPA*. November 2016. https://www.epa.gov/sites/production/files/2016-11/documents/2014_smmfactsheet_508.pdf.

United States Geological Survey. "Water Hardness." Accessed September 18, 2017. https://water.usgs.gov/edu/hardness.html.

The Water Footprint Network. "Product Gallery." Accessed September 17, 2017. http://waterfootprint.org/en/resources/interactive-tools/product-gallery/.

Weikle, Brandie. "Microplastics found in supermarket fish, shellfish." CBC. January 2017. http://www.cbc.ca/news/technology/microplastics-fish-shellfish-1.3954947.

World Wildlife Fund. "Illegal Logging." Accessed September 18, 2017. http://wwf.panda.org/about_our_earth/deforestation/deforestation_causes/illegal_logging/.

ACKNOWLEDGMENTS

First and foremost, I want to thank the incredible and innovative zero waste community. As a blogger and content creator, I am used to publishing my unfinished thoughts and getting instant feedback, so the thought of writing a book admittedly made me slightly anxious.

Without the supportive blog readers of my previous blog on baking, I might not have even considered becoming vegetarian (and then vegan). Without my planet-loving social media followers that readily geeked out with me over worm bins, cloth bags, or uses of stale bread, I might not have pushed things so far. I loved being part of that beehive intelligence and learned so much from everyone every day.

And then I was asked to write a book. Something that would potentially waste paper and other valuable resources, something without a comment section and online notifications. Hoping that a book will be able to reach the people I would not be able to digitally, I took a deep breath and agreed.

The original German version of my book was published in June 2016, and it turned out that my fears were unfounded. In fact, I was blown away! I received an incredible amount of feedback over a diversity of channels, and people did not hesitate to reach out with questions, to share their stories with me, or to just bounce off ideas! I am ever so grateful to be able to contribute to a community I have received so much from.

English is not my native language, and so I would not have been able to translate the book myself without the immense help and edits of fellow eco-nerd, friend, and farmers' market buddy Eszter Lazlo—thank you so much! A big thank you hug also goes to the amazing California-based zero waste blogger Kathryn Kellogg (goingzerowaste.com), who helped me with the localization and patiently put up with my lists of questions and highly random rants. May this book be one of many zero waste books, and may the dream of a zero waste section in every public library come true sooner than later!